Master Math: Algebra 2

Mary Hansen

Cengage Learning PTR

CENGAGE
Learning·

Professional • Technical • Reference

Australia, Brazil, Japan, Korea, Mexico, Singapore, Spain, United Kingdom, United States

CENGAGE
Learning·

Professional • Technical • Reference

**Master Math:
Algebra 2**
Mary Hansen

**Publisher and General
Manager, Cengage Learning
PTR:** Stacy L. Hiquet

**Associate Director of
Marketing:** Sarah Panella

**Manager of Editorial
Services:** Heather Talbot

Senior Marketing Manager:
Mark Hughes

Senior Product Manager:
Emi Smith

Project Editor:
Dan Foster, Scribe Tribe

Technical Editor:
David Lawrence

Interior Layout:
Shawn Morningstar

Cover Designer:
Mike Tanamachi

Proofreader:
Sue Boshers

Printed in the
United States of America
1 2 3 4 5 6 7 16 15 14

For product information and technology
assistance, contact us at
**Cengage Learning Customer and Sales Support,
1-800-354-9706.**

For permission to use material from this
text or product, submit all requests online at
cengage.com/permissions.

Further permissions questions can be
e-mailed to **permissionrequest@cengage.com.**

All trademarks are the property of their respective owners.
All images © Cengage Learning unless otherwise noted.
Library of Congress Control Number: 2013942738
ISBN-13: 978-1-4354-6122-2
ISBN-10: 1-4354-6122-3

Cengage Learning PTR
20 Channel Center Street
Boston, MA 02210
USA

Cengage Learning is a leading provider of customized
learning solutions with office locations around the
globe, including Singapore, the United Kingdom,
Australia, Mexico, Brazil, and Japan. Locate your
local office at: **international.cengage.com/region.**

Cengage Learning products are represented in Canada
by Nelson Education, Ltd.

For your lifelong learning solutions, visit **cengageptr.com.**
Visit our corporate website at **cengage.com.**

To my loving husband.
Thank you for supporting me in every endeavor.

About the Author

Mary Hansen has taught K-12 and post-secondary mathematics and special education in four states. She has travelled the United States extensively, conducting teacher workshops on effective teaching strategies and effective mathematics teaching. Hansen received a Master of Arts in Teaching and a Bachelor of Arts in Mathematics from Trinity University in San Antonio, Texas. She is the co-author of *South-Western Algebra 1: An Integrated Approach*, *South-Western Geometry: An Integrated Approach*, and *South-Western Algebra 2: An Integrated Approach* and author of *Business Math, 17th Edition* and *Master Math: Business and Personal Finance Math.*

Table of Contents

2.4   Scatter Plots and Lines of Regression 42

2.5   Graphing Linear Inequalities 46

Chapter 3 Systems of Equations and Inequalities. 49

3.1   Solving Systems of Equations 50

      Solving a System of Equations by Graphing 50

      Solving a System of Equations by Substitution 52

      Solving a System of Equations by Elimination 54

      Systems with No Solution or Infinite Solutions. 55

3.2   Solving Systems of Inequalities by Graphing 57

3.3   Linear Programming. 62

3.4   Systems of Equations in Three Variables 65

      One Solution. 66

      No Solution or Infinitely Many Solutions 68

3.5   Solving Systems of Equations with Matrices. 71

3.6   Solving Systems of Equations with Determinants
      and Cramer's Rule 74

Chapter 4 Relations and Functions. 81

4.1   Definitions 82

      Relation. 82

      Domain and Range 82

      Functions 84

4.2   Function Notation. 86

4.3   Piecewise Functions 89

      Absolute Value Functions. 91

      Step Functions 92

4.4   Parent Functions and Transformations 92

      Linear Functions. 93

      Absolute Value and Quadratic Functions 94

Introduction

Many believe that there are two kinds of people in this world: those who can do math, and those who can't. I respectfully disagree. While there are certainly people who have a talent or a gift for mathematics, all people are capable of learning and applying mathematics.

In our society, avoiding mathematics is not only a difficult task, but also a risky one. Mathematics is a gatekeeper. If you are unwilling to progress forward, careers and opportunities will be unavailable to you; the gate will be closed.

It has always been my goal as a teacher to make mathematics relevant and understandable to students. While I wish I could sit beside you as you work through this content, I have done my best to explain the topics in everyday language and to show shortcuts and tips that will help you understand the process and perform the mathematics quicker and easier.

What You'll Find in This Book

Master Math: Algebra 2 teaches Algebra 2 from a functions based approach. Essential skills from Algebra 1 will be reviewed as necessary.

You will find all of the major topics covered in an Algebra 2 class:

    Linear Equations and Inequalities

    Quadratic Functions

    Polynomial Functions

    Radical Functions

    Exponential Functions

    Logarithmic Functions

    Rational Functions

In addition, you will find advanced topics:

Sequences and Series

Conic Sections (This bonus Chapter 11 is available for download from the companion website: www.cengageptr.com/downloads.)

Who This Book Is For

Master Math: Algebra 2 is designed as a reference and resource tool. It might be used by a student taking an Algebra 2 or College Algebra course who wants another resource to supplement a textbook, or a parent who needs assistance to help a student through the course. It might be used to refresh skills in preparation for a higher level mathematics course.

How This Book Is Organized

The first three chapters of this book focus on linear equations and inequalities. Topics and techniques from Algebra 1 will be reviewed and expanded.

The fourth chapter is a bridge between linear and more complex equations. The concept of linear and nonlinear functions is introduced, setting the stage to delve into the advanced algebra world of quadratic, polynomial, radical, exponential, logarithmic, and rational functions. The final chapter covers the advanced content of sequences and series. A bonus chapter covering conic sections is available for download from the companion website: www.cengageptr.com/downloads.

Each chapter is broken into named sections so that you can find specific topics easily. Each section has one or more example problems along with several practice problems so that you can do the math to test your skills. An appendix provides not only the answer for each practice problem but also the mathematical steps used to arrive at the answer, so you can check your work each step of the way.

Chapter 1

Linear Equations and Inequalities

1.1 Properties of Real Numbers

Real numbers are the set of numbers made up of all rational and irrational numbers. Subsets of rational numbers include integers, whole numbers, and natural numbers.

- *Natural Numbers*: Often called "counting numbers" {1, 2, 3, 4, ...}.

- *Whole Numbers*: Natural numbers and 0 {0, 1, 2, 3, 4, ...}.

- *Integers*: Whole numbers and their negatives {..., –3, –2, –1, 0, 1, 2, 3, ...}.

- *Rational Numbers*: Numbers that can be written as a ratio of two integers and the denominator of the ratio is not 0, such as
$$\frac{1}{2}, \ -\frac{5}{19}, \ \frac{8}{1}.$$

- *Irrational Numbers*: Numbers that cannot be written as a ratio of two integers, such as π, $\sqrt{2}$, $-\sqrt{5}$. Any non-terminating, non-repeating decimal number and the square root of any number that is not a perfect square are irrational numbers.

The following graphic displays the relationships among the subsets of real numbers:

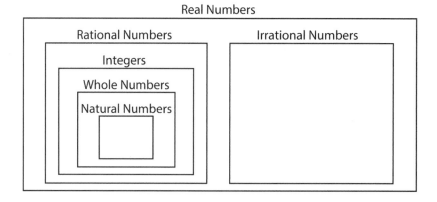

Order of Operations

The *order of operations* establishes the order in which to perform mathematic operations. The importance of the order of operations can be seen in even simple math problems such as these that are routinely posted on Facebook with a challenge to calculate the correct answer:

$$1 + 1 + 1 + 1 + 1 + 1 + 1 + 1 + 1 \cdot 0 =$$

$$6 \div 2(1 + 2) =$$

By the variety of answers people vote for on these kinds of problems, you can see that many people are tripped up by the order of operations!

Order of Operations

- Perform operations within parentheses and other grouping symbols, beginning with the innermost.
- Perform exponent and root calculations.
- Multiply and divide in order from left to right.
- Add and subtract in order from left to right.

An easy way to remember the order of operations is the phrase

"Please Excuse My Dear Aunt Sally"

Please: Parentheses (or other grouping symbols)
Excuse: Exponents
My Dear: Multiplication and Division, left to right
Aunt Sally: Addition and Subtraction, left to right

Using the order of operations,

$$1 + 1 + 1 + 1 + 1 + 1 + 1 + 1 + 1 \cdot 0 = 8$$

since you multiply first and then add.

Example: Evaluate: $[3+(5-2)^3] \cdot 2 + 2 \div 2 - 1$

    Solution: Use the order of operations.

    $[3+(5-2)^3] \cdot 2 + 2 \div 2 - 1$

    $[3+(3)^3] \cdot 2 + 2 \div 2 - 1$ Evaluate $5 - 2$.

    $[30] \cdot 2 + 2 \div 2 - 1$ Evaluate 3^3 then add 3.

    $60 + 1 - 1$ Multiply and divide left to right.

    60 Add and subtract left to right.

Practice Exercises

1.1 Evaluate: $6 \div 2(1+2)$

1.2 Evaluate: $12 \div 4 + 2 \cdot 6 - (4-3)^3$

Properties of Addition and Multiplication

A *property* is an attribute of something. Mathematical properties are statements that are true for all items or numbers in the set.

For all real numbers a, b, and c,

Closure Property:

    Addition $a + b$ is a real number

    Multiplication $a \cdot b$ is a real number

Commutative Property:

    Addition $a + b = b + a$

    Multiplication $a \cdot b = b \cdot a$

Associative Property:

    Addition $a + (b + c) = (a + b) + c$

    Multiplication $a \cdot (b \cdot c) = (a \cdot b) \cdot c$

Identity Property:

    Addition $a + 0 = a$

    Multiplication $a \cdot 1 = a$

Inverse Property:

Addition $\qquad a + (-a) = 0$

Multiplication $\qquad \dfrac{1}{a} \cdot a = 1,\ a \neq 0$

Property Hints

- Closure indicates that if you start with two numbers in a set, and you perform an operation, the result will also be in the set.

- Commutative means that you can reverse the order of the operation. For example, $4 + 3 = 3 + 4$. Note that subtraction and division are not commutative since $6 - 2 \neq 2 - 6$ and $6 \div 2 \neq 2 \div 6$.

- Associative means that you can change the grouping or "associations" of the numbers while maintaining the same order.

- The identity is a number that, for a certain operation, gives back the original number.

- Performing the operation on a number and its *inverse* produces the identity.

- These properties are specific to the set of numbers and a specific operation.

Example: Name the property illustrated.

(a) $2 + (4 + 5) = (2 + 4) + 5$

(b) $6 + (-6) = 0$

(c) $5(7) = 7(5)$

Solution:

(a) The order of the numbers is the same, only the grouping has changed. Associative Property of Addition.

(b) The numbers being added are inverses because the result is the identity for addition. Inverse Property of Addition.

(c) The order of the factors has been reversed. Commutative Property of Multiplication.

Practice Exercises

Name the property illustrated.

1.3 $3(1) = 3$

1.4 $2x + 1 = 1 + 2x$

1.5 $3 \cdot (-5) = -5 \cdot 3$

Distributive Property

The properties in the previous section applied to single operations, either multiplication or addition. The *Distributive Property of Multiplication Over Addition* relates both operations.

Distributive Property of Multiplication Over Addition

$$a(b + c) = ab + ac$$

The Distributive Property can help simplify calculations.

Example: Simplify: $6(10 + 8)$

   Solution: Using the order of operations you add first, then multiply.

   $6(10 + 8) = 6(18) = 108$

   Using the Distributive Property, and then the order of operations, you distribute and multiply first, then add.

   $6(10 + 8) = 6(10) + 6(8) = 60 + 48 = 108$

   Multiplying 6 by 10 and 8 separately and then adding is easier for most people to do in their head than multiplying 6 by 18.

The Distributive Property can even help you simplify calculations like figuring a tip.

Example: Christina wants to leave a 15% tip on a $30 restaurant bill. How much should the tip be?

   Solution: Multiply $30 by 15%.

   $30(15\%) = \$4.50$

   Use the Distributive Property to make the problem easier to calculate.

   $30(15\%) = \$30(10\% + 5\%) = \$30(10\%) + \$30(5\%)$

To multiply by 10%, move the decimal point in the $30 to the left one place, so that $30 becomes $3. Five percent is half of 10%, and half of $3 is $1.50.

$30(15\%) = \$30(10\% + 5\%) = \$30(10\%) + \$30(5\%) = \$3 + \$1.50 = \$4.50$

Using the Distributive Property in reverse can help you simplify other problems.

Example: Simplify: 8(6) + 8(14)

Solution: Using the order of operations, you multiply first, and then add.

$8(6) + 8(14) = 48 + 112 = 160$

Using the Distributive Property and then the order of operations, you add then multiply.

$8(6) + 8(14) = 8(6 + 14) = 8(20) = 160$

Practice Exercises

Use the Distributive Property to help simplify.

1.6   8(20 + 3)

1.7   9(14) + 9(16)

1.2 Equations

An *equation* says that two things have equal value. For example, $3 + 7 = 12 - 2$ is an equation, as is $2x - 3 = 4$.

$2x - 3 = 4$ 2 is a *coefficient*

3 and 4 are *constants*

x is a *variable* or an *unknown*

At its simplest level, *solving an equation* that has a variable means to find the value or values for the variable that make the equation true. In the equation above, $x = 3.5$ because $2(3.5) - 3 = 4$.

Properties of Equality

Recall that a property is a statement that is true for all numbers or items in the set. Properties of Equality are helpful in solving equations.

For all real numbers a, b, and c, the Properties of Equality are:

Reflexive Property	$a = a$
Symmetric Property	If $a = b$, then $b = a$
Transitive Property	If $a = b$ and $b = c$, then $a = c$
Substitution Property	If $a = b$, then b can replace a or a can replace b
Addition Property	If $a = b$, then $a + c = b + c$
Subtraction Property	If $a = b$, then $a - c = b - c$
Multiplication Property	If $a = b$, then $ac = bc$
Division Property	If $a = b$ and $c \neq 0$, then $\dfrac{a}{c} = \dfrac{b}{c}$

Whether or not you realize it, when you solve an equation by writing equivalent equations, you are using the Properties of Equality.

Property Hints

- *Reflexive* is like a reflection in the mirror; a variable or number has the same value as itself.

- Things that are *symmetric* are the same on both sides of the line of symmetry, just like the value of variables or numbers on each side of an equals sign.

- *Transitive* is like a train. If a is connected to b, and b is connected to c, then a and c are connected.

- The *Addition, Subtraction, Multiplication,* and *Division Properties of Equality* all say that two sides of an equation remain equal if you perform the same operation to each side of the equation, as long as you do not divide both sides of the equation by zero. (Remember that division by zero is undefined.)

Example: Identify which Properties of Equality are being used to solve the equation.

$$2x - 3 = 4$$
$$\text{(a) } 2x - 3 + 3 = 4 + 3$$
$$2x = 7$$

$$\text{(b) } \frac{2x}{2} = \frac{7}{2}$$

$$x = 3.5$$

Solution: In (a), 3 is being added to both sides: Addition Property of Equality. In (b), both sides are being divided by 2: Division Property of Equality.

Notice in the previous example that the five equations listed are all *equivalent equations*. In other words, you can substitute 3.5 for x in any of the three equations and the equation will be true.

In this book, the steps that show the Properties of Equality being applied will not generally be written out. So the solution to the equation $2x - 3 = 4$ would be written as:

$$2x - 3 = 4$$
$$2x = 7 \qquad \text{Add 3 to both sides.}$$
$$x = 3.5 \qquad \text{Divide both sides by 2.}$$

Practice Exercises

Identify the property used to transform the equation into an equivalent equation.

$$\frac{2}{3}x + 4 = 8$$

1.8 $\frac{2}{3}x = 4$

1.9 $x = 6$

Solving Equations

Linear equations can look simple, such as $\frac{2}{5}x - 1 = 5$, or more complex,

such as $6x - 6 = -\frac{2}{3}(6x - 45) - 2x$. A linear equation is an equation

that contains only constants or the product of a number and a variable to the first power.

Example: Solve: $\frac{2}{5}x - 1 = 5$

   Solution:

$$\frac{2}{5}x - 1 = 5$$

$$\frac{2}{5}x = 6 \qquad \text{Add 1 to both sides.}$$

$$x = \overset{3}{\cancel{6}}\left(\frac{5}{\underset{1}{\cancel{2}}}\right) \qquad \text{Multiply both sides by } \frac{5}{2}, \text{ the inverse of } \frac{2}{5}.$$

$$x = 15$$

Substitute 15 for x in the original equation to check your work.

$$\frac{2}{5}(15) - 1 = 5$$

$$5 = 5 \checkmark$$

Using simplification, the Distributive Property, and Properties of Equality, you can simplify a complex equation. Remember that terms on the same side of the equation that have the same variable can be combined. $6x + 3x = 9x$ and $2x - 5x = -3x$ are examples of combining like terms. However, recall that variables that don't have the same exponent cannot be combined. x^2 and x cannot be combined.

Example: Solve: $6x - 6 = -\dfrac{2}{3}(6x - 45) - 2x$

Solution:

$$6x - 6 = -\frac{2}{3}(6x - 45) - 2x$$

$6x - 6 = -4x + 30 - 2x$	Use the Distributive Property.
$6x - 6 = -6x + 30$	Combine like terms.
$12x - 6 = 30$	Add $6x$ to both sides.
$12x = 36$	Add 6 to both sides.
$x = 3$	Divide both sides by 12.

Check:

$$6(3) - 6 = -\frac{2}{3}(6 \cdot 3 - 45) - 2(3)$$

$$12 = 12 \checkmark$$

In the previous problem, the fraction distributed nicely into the parentheses and no fractions remained. However, often it is easier to get rid of the fractions by multiplying both sides of the equation by the least common multiple of the denominators of the fractions.

Example: Solve: $\dfrac{1}{2}(2x + 5) = \dfrac{2}{3}(9x - 7)$

Solution:

$$\frac{1}{2}(2x + 5) = \frac{2}{3}(9x - 7)$$

$$\overset{3}{\cancel{6}}\left(\frac{1}{\underset{1}{\cancel{2}}}(2x + 5)\right) = \overset{2}{\cancel{6}}\left(\frac{2}{\underset{1}{\cancel{3}}}(9x - 7)\right) \quad \text{Multiply both sides by 6.}$$

$3(2x + 5) = 4(9x - 7)$	Simplify.
$6x + 15 = 36x - 28$	Use the Distributive Property.
$15 = 30x - 28$	Subtract $6x$ from both sides.
$43 = 30x$	Add 28 to both sides.
$\dfrac{43}{30} = x$	Divide both sides by 30.

Check:

$$\frac{1}{2}\left(2 \cdot \frac{43}{30} + 5\right) = \frac{2}{3}\left(9 \cdot \frac{43}{30} - 7\right)$$

$$\frac{59}{15} = \frac{59}{15} \checkmark$$

You will make fewer mistakes with positive and negative signs if you move the variable to the side of the equation where it will have a positive coefficient. In the previous example, that meant moving the variable to the right side of the equation.

Even a problem full of fractions can simplify quickly once you clear the fractions.

Example: $\dfrac{3x+1}{6} + \dfrac{-x+2}{9} = \dfrac{2x+1}{4}$

Solution:

$$36\left(\frac{3x+1}{6} + \frac{-x+2}{9}\right) = 36\left(\frac{2x+1}{4}\right) \qquad \text{Multiply both sides by 36.}$$

$$36\left(\frac{3x+1}{6}\right) + 36\left(\frac{-x+2}{9}\right) = 36\left(\frac{2x+1}{4}\right) \qquad \begin{array}{l}\text{Use the Distributive}\\ \text{Property.}\end{array}$$

$$\overset{6}{\cancel{36}}\left(\frac{3x+1}{\cancel{6}}\right) + \overset{4}{\cancel{36}}\left(\frac{-x+2}{\cancel{9}}\right) = \overset{9}{\cancel{36}}\left(\frac{2x+1}{\cancel{4}}\right) \qquad \text{Simplify.}$$

$$6(3x + 1) + 4(-x + 2) = 9(2x + 1) \qquad \text{Simplify.}$$

$$18x + 6 - 4x + 8 = 18x + 9 \qquad \begin{array}{l}\text{Use the Distributive}\\ \text{Property.}\end{array}$$

$$14x + 14 = 18x + 9 \qquad \text{Simplify.}$$

$$14 = 4x + 9 \qquad \begin{array}{l}\text{Subtract } 14x \text{ from both}\\ \text{sides.}\end{array}$$

$$5 = 4x \qquad \text{Subtract 9 from both sides.}$$

$$x = \frac{5}{4} \qquad \text{Divide both sides by 4.}$$

The check is left to you.

General Tips for Solving Linear Equations

- Clear fractions by multiplying both sides of the equation by a number that will "cancel out" the denominators of the fractions.

- Simplify each side of the equation by using tools such as the Distributive Property and combining like terms.

- Use the Addition or Subtraction Property of Equality to put the variables on the same side of the equation.

- Use the Properties of Equality to solve for the variable.

- You have a solution when you have isolated the variable on one side of the equation, and its coefficient is 1.

- Check your solution by substituting it into the original equation and verifying that it makes the equation true.

One last hint: If you solve a problem and the variable drops out, leaving you with a statement like $4 = 6$ or $0 = 0$, then either your original equation has no solution, or all real numbers are solutions. How do you decide? If the statement is true, then all real numbers are the solution to the equation. If the statement is false, then no real numbers are a solution.

$$x + 6 = x + 6 \qquad\qquad x + 6 = x + 7$$
$$6 = 6 \qquad\qquad\qquad 6 \neq 7$$

All real numbers   No real numbers, No solution

In the equation $x + 6 = x + 6$, you can substitute any real number for x and get a true statement. In the equation $x + 6 = x + 7$, there are no real numbers that you can substitute for x and get a true statement.

Practice Exercises

Solve.

1.10 $5x - 3 = 6 + 3x$

1.11 $\dfrac{2}{5}(4x - 1) = -\dfrac{3}{10}(4 - 3x)$

Sometimes you will work with an equation that has more than one variable, and you will be asked to solve for a certain variable. In other words, you are to isolate that variable on one side of the equation. In this case, your solution will most likely not be a single value but an expression that contains one or more variables.

Example: The formula to convert Celsius to Fahrenheit is $F = \dfrac{9}{5}C + 32$ where C is the temperature in Celsius, and F is the temperature in Fahrenheit. Solve the formula for C.

Solution:

$$F = \frac{9}{5}C + 32$$

$$F - 32 = \frac{9}{5}C \quad \text{Subtract 32 from both sides.}$$

$$\frac{5}{9}(F - 32) = C \quad \text{Multiply both sides by } \frac{5}{9}.$$

Practice Exercises

1.12 $R = DL + S$; solve for D

1.13 $3x + 4y = 12$; solve for y

1.3 Absolute Value Equations

The *absolute value* of a number is the distance of the number from zero. For example, because negative six and positive six are both six units away from zero, the absolute value of both numbers is six. Some people describe the absolute value of a number as its numerical value without regard to sign.

The absolute value of a number is designated by two vertical bars. $|-6| = 6$ and $|6| = 6$.

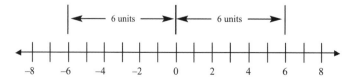

In general notation, for any real number a,

$|a| = a$ if $a \geq 0$ and $|a| = -a$ if $a < 0$

While the previous notation may look confusing, all it says is that if a number is 0 or greater, the absolute value is the same as the number itself. If the number is less than 0, or negative, then the absolute value is the opposite of the number, so the absolute value of the number is positive.

If $|x| = 3$, then $x = 3$ or $x = -3$.

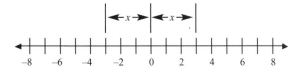

To solve an absolute value equation:

• Isolate the absolute value expression.

• Remove the absolute value symbol and write two equations: one that is the same as the original equation without the absolute value symbol, and one that is equal to its opposite.

• Connect the two equations with "or."

• Solve each equation.

• Check the values found. Values that do not make the original equation true are called *extraneous solutions*, and are not included as solutions.

Example: $|3x - 1| = 5$

   Solution: $3x - 1$ is 5 units away from 0.

$$3x - 1 = 5 \quad \text{or} \quad 3x - 1 = -5$$
$$3x = 6 \quad \text{or} \quad 3x = -4$$
$$x = 2 \quad \text{or} \quad x = -\frac{4}{3}$$

So x could be 2 or $-\dfrac{4}{3}$. Check each value in the original equation.

Check $x = 2$.   Check $x = -\dfrac{4}{3}$.

$|3(2) - 1| = 5$ $\left|3\left(-\dfrac{4}{3}\right) - 1\right| = 5$

$\qquad |5| = 5$ $\qquad |-5| = 5$

$\qquad 5 = 5\ \checkmark$ $\qquad 5 = 5\ \checkmark$

$x = 2,\ -\dfrac{4}{3}$

Remember to isolate the absolute value expression before you split the equation into two equations.

Example: $\dfrac{1}{2}\left(|x + 5| - 2\right) = -x$

Solution:

$\dfrac{1}{2}\left(|x + 5| - 2\right) = -x$

$|x + 5| - 2 = -2x$	Multiply both sides by 2.
$\qquad |x + 5| = -2x + 2$	Add 2 to both sides.
$x + 5 = -2x + 2$ or $x + 5 = -(-2x + 2)$	Write two equations.
$x + 5 = -2x + 2$ or $x + 5 = 2x - 2$	Use the Distributive Property.
$3x + 5 = 2$ or $5 = x - 2$	Move variables to the same side.
$3x = -3$ or $7 = x$	Add/subtract from both sides.
$x = -1$ or $7 = x$	Solve for x.

Check $x = -1$.   Check $x = 7$.

$\dfrac{1}{2}\left(|-1 + 5| - 2\right) = -(-1)$ $\dfrac{1}{2}\left(|7 + 5| - 2\right) = -(7)$

$\qquad\qquad 1 = 1\ \checkmark$ $\qquad\qquad 5 = -7$ FALSE

So -1 is a solution, and 7 is an extraneous solution.

$x = -1$

Watch out for problems where the absolute value expression is equal to a negative number, such as $|x| = -5$ or $|2x + 1| = -3$. These equations have *no solution*. Remember that absolute value is a distance, and the measurement of distance is greater than or equal to 0.

Practice Exercises

1.14 $|3x - 2| = 4$

1.15 $3|2x + 7| = 3x - 6$

1.4 Inequalities

While an equation states that two expressions are equal, an *inequality* states that one expression is less than or greater than another expression. The following are all examples of inequalities.

$x < 0$ x is *less than* 0.

$x \leq 3$ x is *less than or equal to* 3.

$x > -1$ x is *greater than* -1.

$x \geq -4.2$ x is *greater than or equal to* -4.2.

Solving Inequalities

You can use the same techniques to solve inequalities as you did with equations, with two exceptions. If you multiply or divide both sides of the inequality by a negative number, then you must reverse the inequality symbol; less than becomes greater than, and greater than becomes less than.

For all real numbers a, b, and c, the Properties of Inequality are:

Addition Property   If $a < b$, then $a + c < b + c$

Subtraction Property   If $a < b$, then $a - c < b - c$

Multiplication Property   If $a < b$, then $ac < bc$ if $c > 0$
$$ac > bc \text{ if } c < 0$$

Division Property   If $a < b$, then $\dfrac{a}{c} < \dfrac{b}{c}$ if $c > 0$

$$\dfrac{a}{c} > \dfrac{b}{c} \text{ if } c < 0$$

Remember to reverse the direction of the inequality symbol if you multiply or divide both sides of an inequality by a negative number!

Example: $-3x + 2 < 2x + 12$

Solution:

$-3x + 2 < 2x + 12$

$-5x + 2 < 12$ Subtract $2x$ from both sides.

$-5x < 10$ Subtract 2 from both sides.

$x > -2$ Divide both sides by -5 and reverse inequality.

Another way to solve the inequality is to solve so that the coefficient of x remains positive.

$-3x + 2 < 2x + 12$

$2 < 5x + 12$ Add $3x$ to both sides.

$-10 < 5x$ Subtract 12 from both sides.

$-2 < x$ Divide both sides by 5.

This solution says -2 is less than x, which is another way to say that x is greater than -2.

Both methods give the same solution: $x > -2$.

Substitute a value from the solution set into the original inequality to check your work.

Check $x = 0$.

$-3(0) + 2 < 2(0) + 12$

$2 < 12$ ✓

Graphing Solutions to Inequalities

Solutions to inequalities are often graphed on a number line to give a visual picture of the solutions.

$x < 3$ $x \leq 3$ $x > 3$ $x \geq 3$

A *solid dot* is used with $\leq$ or $\geq$ to show that the point is *included* in the solution.

An *open dot* is used with $<$ or $>$ to show that the point is *not included* in the solution.

Example: Solve and graph the solution: $-6x + 4 > 28$.

> **Solution:**
>
> $-6x + 4 > 28$
>
> $\quad -6x > 24$ Subtract 4 from both sides.
>
> $\quad\quad x < -4$ Divide both sides by -6 and reverse inequality.
>
> The solution is all real numbers less than -4, but not including -4.

You can check your work by substituting in any number that is shaded on the graph.

Check $x = -6$.

$-6(-6) + 4 > 28$

$\quad\quad 40 > 28 \checkmark$

Practice Exercises

Solve and graph the solution.

1.16 $9x - 3 < -5x + 4$

1.17 $-2(x + 5) \geq \dfrac{1}{2}(5x - 2)$

1.5 Compound Inequalities

A *compound statement* is a statement made up of two ideas joined by *and* or *or*. We use compound statements all the time. "I'll call you before six or after eight." "After six and before eight, I'll be watching a movie." Or, a more common statement might be, "Between six and eight, I'll be watching a movie." *Between* is a shortcut for two statements joined by *and*.

In our everyday speech, we understand that a compound statement with *or* means at least one of the statements is (or will be) true, and we understand that a compound statement with *and* means that both statements are (or will be) true.

Mathematically, these statements can be represented with *compound inequalities*:

"Before six or after eight"	$x < 6$ or $x > 8$	disjunction
"After six and before eight"	$x > 6$ and $x < 8$	conjunction
"Between six and eight"	$6 < x < 8$	conjunction

Notice that:

"After six and before eight" is equivalent to "Between six and eight," just as $x > 6$ and $x < 8$ is equivalent to $6 < x < 8$.

A compound inequality joined by *and* is called a *conjunction*. A compound inequality joined by *or* is called a *disjunction*.

To be a solution for an *and* compound inequality (conjunction), a value must satisfy *both* parts of the inequality.

To be a solution for an *or* compound inequality (disjunction), a value must satisfy *at least one* part of the inequality.

Example: Solve and graph the solution to the compound inequality.

$$\frac{1}{3}(2x-1) > 7 \quad \text{or} \quad -2x + 7 \le -3x + 5$$

Solution:

$$\frac{1}{3}(2x-1) > 7 \qquad \text{or} \qquad -2x + 7 \le -3x + 5$$

$$2x - 1 > 21 \quad \text{or} \qquad x + 7 \le 5$$

$$2x > 22 \quad \text{or} \qquad x \le -2$$

$$x > 11 \quad \text{or} \qquad x \le -2$$

x is greater than 11 or less than or equal to –2.

It is advisable to check solutions from both parts of the graph.

Check $x = 14$.   Check $x = -3$.

$$\frac{1}{3}(2 \cdot 14 - 1) > 7 \qquad\qquad -2(-3) + 7 \le -3(-3) + 5$$

$$9 > 7 \checkmark \qquad\qquad\qquad 13 \le 14 \checkmark$$

The graph of the solution to a conjunction is usually a graph with one part, so you only need to check a single value, but it should be checked in both inequalities to verify that it makes a true statement.

Example: Solve and graph the solution to the compound inequality.

$$2x \ge -\frac{1}{2}(10+x) \quad \text{and} \quad 2(x-1) < -(x+2)$$

Solution:

$$2x \ge -\frac{1}{2}(10+x) \quad \text{and} \quad 2(x-1) < -(x+2)$$

$$-4x \le 10 + x \qquad \text{and} \qquad 2x - 2 < -x - 2$$
$$-5x \le 10 \qquad\quad \text{and} \qquad 3x - 2 < -2$$
$$x \ge -2 \qquad\quad \text{and} \qquad\quad 3x < 0$$
$$x \ge -2 \qquad\quad \text{and} \qquad\quad x < 0$$

x is greater than or equal to -2 and less than 0.

Check $x = -1$.

$$2(-1) \ge -\frac{1}{2}(10 + (-1)) \quad \text{and} \quad 2(-1-1) < -(-1+2)$$

$$-2 \ge -\frac{9}{2} \checkmark \qquad\qquad \text{and} \qquad\qquad -4 < -1 \checkmark$$

Remember that when a compound inequality is written as a combined statement, it is a conjunction, or an *and* statement. You can break up the combined statement, or you can solve it as it is written. If you solve a combined inequality as written, you must apply the Properties of Inequality to all three expressions.

Example: Solve and graph the solution to the compound inequality.

$$0 < \frac{2}{3}(x+1) < 4$$

Solution:

$$0 < \frac{2}{3}(x+1) < 4$$

$0 < 2(x + 1) < 12$	Multiply all expressions by 3.
$0 < 2x + 2 < 12$	Use the Distributive Property.
$-2 < \quad 2x \quad < 10$	Subtract 2 from all expressions.
$-1 < \quad x \quad < 5$	Divide all expressions by 2.

x is between -1 and 5.

x is greater than -1 and less than 5.

Performing the check is left to you.

In the previous example, the combined inequality could have been split into two inequalities joined by "and": $0 < \frac{2}{3}(x+1)$ and $\frac{2}{3}(x+1) < 4$.

Watch out for problems where the solution is *all real numbers* or *no solution*.

$x > 4$ or $x < 6$ All real numbers are greater than 4 or less than 6.

$x < 4$ and $x > 6$ No real numbers are both less than 4 and greater than 6.

Practice Exercises

Solve and graph.

1.18 $\dfrac{7x+11}{3} \geq 6$ or $3(-2x + 1) > 15$

1.19 $3x + 1 > 5$ and $-2x + 6 > 4$

1.20 $3 < \dfrac{1}{3}(x+5) < 4$

1.6 Absolute Value Inequalities

An absolute value inequality that uses *less than*, such as $|x| < 6$, is saying that x is less than 6 units away from zero. So, x must be greater than -6 and less than 6. Using symbols, $x > -6$ and $x < 6$. You can also say that x is between -6 and 6 and use the combined inequality $-6 < x < 6$.

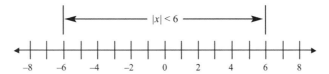

An absolute value inequality that uses *greater than*, such as $|x| > 6$, is saying that x is more than 6 units away from zero. So, x must be less than -6 or x must be greater than 6. Using symbols, $x < -6$ or $x > 6$.

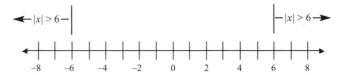

To solve an absolute value inequality, isolate the absolute value expression and then rewrite the inequality as a compound inequality. When the inequality is "less than" or "less than or equal to," use a conjunction—an *and* statement. When the inequality is "greater than" or "greater than or equal to," use a disjunction—an *or* statement.

Isolate the absolute value expression.

Rewrite the absolute value inequality as a compound inequality.

$<, \leq$ Inequalities joined by and.

$>, \geq$ Inequalities joined by or.

Solve the inequalities and then check and graph the solution.

Example: Solve and graph the solution: $|2x - 1| < 9$

> **Solution:** The absolute value is less than 9, so it is greater than -9 and less than 9.

$2x - 1 > -9$ and $2x - 1 < 9$ Write two inequalities joined by and.

$2x > -8$ and $2x < 10$ Add 1 to both sides.

$x > -4$ and $x < 5$ Divide both sides by 2.

x is greater than -4 and less than 5.

Another way to solve the problem is to use a combined inequality:

$-9 < 2x - 1 < 9$

$-8 <$ $2x$ < 10 Add 1 to all expressions.

$-4 <$ x < 5 Divide all expressions by 2.

x is between -4 and 5.

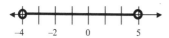

Check your work by substituting in any shaded number on the graph.

Check $x = 0$.

$|2(0) - 1| < 9$

$1 < 9$ ✓

You should always check your answer. Choose one number or several numbers from your solution, substitute them into the original inequality, and make sure that they produce a true statement.

Just like when you solve absolute value equations, you should isolate the absolute value expression before you split the inequality into two inequalities.

Example: Solve and graph the solution: $2|-x + 4| > 12$

Solution:

$2|-x + 4| > 12$

$|-x + 4| > 6$ Isolate the absolute value expression.

The absolute value is greater than 6, so it is less than -6 or greater than 6.

$-x + 4 < -6$ or $-x + 4 > 6$ Write two inequalities with or.

$-x < -10$ or $-x > 2$ Subtract 4 from both sides.

  $x > 10$ or $x < -2$ Divide both sides by -1; switch inequalities.

x is greater than 10 or less than -2.

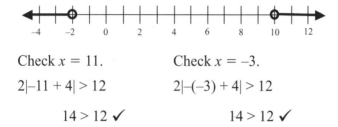

Check $x = 11$.   Check $x = -3$.

$2|{-}11 + 4| > 12$ $2|{-}(-3) + 4| > 12$

       $14 > 12$ ✓ $14 > 12$ ✓

Watch out for absolute value expressions that are less than 0 or a negative number, such as $|x| < 0$ or $|3x + 4| \le -2$. These inequalities have *no solution*. Remember that absolute value is a measure of distance, and a distance cannot be less than 0.

Watch out for absolute value expressions that are greater than or equal to 0 or greater than a negative number, such as $|x| \ge 0$ or $|3x + 4| > -2$. The solution to these inequalities is *all real numbers*. Since absolute value is a measure of distance, any distance will be greater than or equal to 0 or greater than a negative number.

Practice Exercises

Solve and graph the solution.

1.21 $|3x - 5| < 4$

1.22 $-\dfrac{1}{2}|4x + 8| < 6$

Chapter 2

Graphs of Linear Equations and Inequalities

2.1 Graphing Equations

In Chapter 1, you solved equations and inequalities in one variable and illustrated the solutions on a number line. An equation or inequality in two variables often uses the variables x and y, and the solutions are the ordered pairs (x, y) that make the equation or inequality true. The ordered-pair solutions can be graphed on a coordinate plane.

Linear vs. Nonlinear

A *linear equation* is an equation whose solutions form a straight line on a graph. An equation that produces a graph that is not a straight line is *nonlinear*.

Example: Graph the equation $y = x^2$. Is the equation linear or nonlinear?

> **Solution:** Choose several different values for x, substitute each value into the equation, and find the value of y. Graph the ordered pairs and connect them with a line or smooth curve.

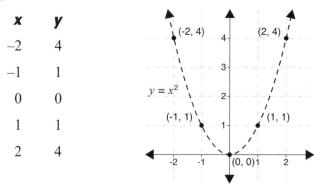

x	y
–2	4
–1	1
0	0
1	1
2	4

The graph is a curve, so the equation is nonlinear.

If a variable in the equation has a fractional coefficient, try to choose values for the variable that will eliminate the fraction and produce integer ordered pairs.

Example: Graph the equation $y = -\dfrac{2}{3}x + 1$. Is the equation linear or nonlinear?

> **Solution:** Choose several different values for x, substitute each value into the equation, and find the values of y. Graph the ordered pairs and connect them with a line or smooth curve.

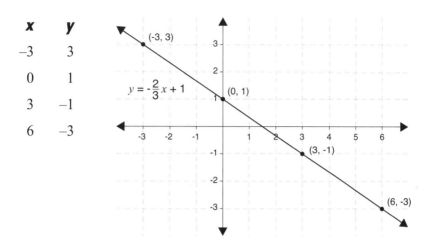

The graph is a straight line, so the equation is linear.

Linear equations will *not* have exponents other than one, variables in the denominator of a fraction, or variables that are multiplied by one another.

Examples of linear equations	Examples of nonlinear equations
$y = -2x + 3$	$y = 3x^2$
$x - y = 6$	$y = \dfrac{1}{x}$
$4x + 7y + 1 = 0$	$xy = 4$

Practice Exercises

Graph the equations.

2.1 $y = 3x + 1$

2.2 $x - y = 6$

Slope-Intercept Form

Graphing a linear equation by choosing values and solving for the other variable is not a very efficient way to graph. Instead, you can use information given by different forms of a linear equation to draw the graph.

A linear equation can be written in *slope-intercept form.*

$$y = mx + b$$

In slope-intercept form, m and b are constants, where m represents the slope of the line and b represents the y-intercept of the line.

The *y-intercept* of the line is the place where the line crosses the y-axis.

The *slope* of a line can be described as the "steepness" of a line, or how much vertical change there is for a unit of horizontal change. Slope can also be called the *rate of change.*

The graphs in Figure 2.1 illustrate the intuitive notion of rate of change or slope. Each graph represents the height of water in a bathtub over time.

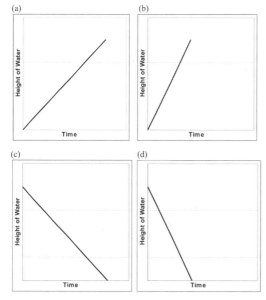

Figure 2.1
Rate of change, or slope

Graphs (a) and (b) represent a bathtub being filled with water. As time passes, the height of the water increases. Notice that the height of the water increases faster in graph (b). Both graphs represent a positive rate of change, or positive slope, but graph (b) shows a faster rate of change.

Graphs (c) and (d) represent a bathtub being drained of water. As time passes, the height of the water decreases. Notice that the height of the water decreases faster in graph (d). Both graphs represent a negative rate of change, or negative slope, but graph (d) shows a faster rate of change.

On a standard graph, the slope of a line can be counted from the graph or calculated from two points on the graph. To count the slope, start at one point on the line and count the *rise*, the number of units up or down, and the *run*, the number of units left or right, to travel to the other point. The slope is equal to the rise over the run.

$$Slope = \frac{rise}{run}$$

Up is positive rise, while down represents negative rise.
Right is positive run, while left represents negative run.

Another way to find slope is to calculate it from two points on the line.

$$Slope = \frac{(y_2 - y_1)}{(x_2 - x_1)}$$

Where (x_1, y_1) and (x_2, y_2) are two points on the line.

Example: Find the slope and the *y*-intercept of the line.

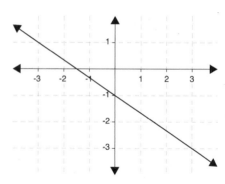

Solution: Identify the point where the graph crosses the *y*-axis. Find the slope by counting the rise and run or calculating it from two points on the line.

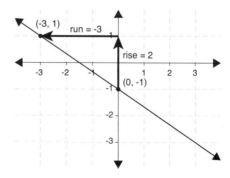

The line crosses the *y*-axis at $(0, -1)$, so the *y*-intercept is -1.

Starting at $(0, -1)$ and counting to $(-3, 1)$, you can count up 2 units and left 3 units. Rise $= 2$, run $= -3$, slope $= -\dfrac{2}{3}$.

You can also calculate the slope by using the points $(0, -1)$ and $(-3, 1)$.

$$\text{Slope} = \frac{(y_2 - y_1)}{(x_2 - x_1)} = \frac{(1 - (-1))}{(-3 - 0)} = \frac{2}{-3} = -\frac{2}{3}$$

The slope of the line is $-\dfrac{2}{3}$ and the y-intercept is -1.

Practice Exercises

2.3   What is the slope and the y-intercept of the line?

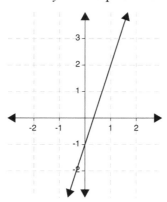

2.4   What is the slope of a line that passes through the points $(2, -5)$ and $(1, 3)$?

The slope-intercept form can be used to graph a linear equation.

Example: Graph the equation $-2x + y = -3$

   Solution: Solve the equation for y. Identify the slope and the y-intercept. Plot the y-intercept and then count the rise and run to find another point. Draw a line through the two points.

   $-2x + y = -3$

   $\qquad y = 2x - 3$ Add $2x$ to both sides.

   $b = -3$, so the y-intercept is -3. Graph the point $(0, -3)$.

The slope is 2, or $\dfrac{rise}{run} = \dfrac{2}{1}$, so from (0, –3), count up 2 and to the

right 1 and plot another point.

Draw a line through the points.

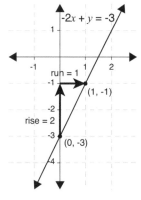

Practice Exercises

Graph.

2.5 $y = -\dfrac{3}{2}x + 2$

2.6 $x - 2y = 6$

Standard Form of a Linear Equation

The *standard form* of a linear equation is $Ax + By = C$, where $A \geq 0$; A, B, and C are integers; and A and B are not both 0.

Example: Write the linear equation $y = \dfrac{2}{5}x - 3$ in standard form and identify the values of A, B, and C.

> **Solution:** Eliminate the fraction by multiplying both sides of the equation by 5. Transform the equation so that x and y are on the left side of the equation.
>
> $$y = \dfrac{2}{5}x - 3$$
>
> $5y = 2x - 15$ Multiply both sides by 5.
>
> $-2x + 5y = -15$ Subtract $2x$ from both sides.
>
> $2x - 5y = 15$ Multiply both sides by -1.
>
> $A = 2, B = -5, C = 15$

Practice Exercises

Write the linear equation in standard form and identify the values of A, B, and C.

2.7 $y = -\dfrac{3}{2}x + 2$

2.8 $y = 6$

You can use the standard form of a linear equation to easily find the x- and y-intercepts and draw the line.

The *intercepts* are the places where the graph crosses the axes. The *x-intercept* is the point at which the graph crosses the x-axis, and the *y-intercept* is the point at which the graph crosses the y-axis.

Example: Graph the linear equation $5x - 2y = 4$

> **Solution:** Substitute $y = 0$ and solve for x to find the x-intercept. Substitute $x = 0$ and solve for y to find the y-intercept.

x-intercept, substitute $y = 0$	y-intercept, substitute $x = 0$
$5x - 2(0) = 4$	$5(0) - 2y = 4$
$5x = 4$ Simplify.	$-2y = 4$ Simplify.
$x = \dfrac{4}{5}$ Divide by 5.	$y = -2$ Divide by -2.
$\left(\dfrac{4}{5}, 0\right)$; $\dfrac{4}{5}$ is the x-intercept	$(0, -2)$; -2 is the y-intercept

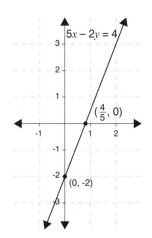

Practice Exercises

Use the x- and y-intercepts to graph the linear equations.

2.9 $2x - y = -5$

2.10 $3x + 2y = 6$

Special Cases: Horizontal and Vertical Lines

An equation that represents a horizontal line has the form $y = 0x + b$, or $y = b$, where b is a constant. The slope of a horizontal line is 0 and the y-intercept is b. In the graph at the right, the horizontal line is represented by the equation $y = 2$. No matter what value is chosen for x, $y = 2$.

A vertical line has the form $x = C$ where C is a constant. The x-intercept is C and the slope is undefined because the run is 0. In other words, x is always equal to C regardless of the value of y. In the graph at the right, the equation of the vertical line is $x = -1$.

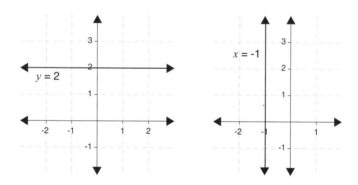

Horizontal lines have the form $y = b$ and a slope of 0. Vertical lines have the form $x = C$ and the slope is undefined.

2.2 Writing Linear Equations

It is important to be able to use given information about a line to write the equation of the line. Depending on what information you know, you will use slope-intercept or point-slope forms of a linear equation to write the equation of a line or the equation of a line that is parallel or perpendicular to a given line.

Slope-Intercept Form

You saw in the previous section that you can use the graph of a line to find the slope and the y-intercept of the line. Knowing the slope and the y-intercept, you can plug the values into the slope-intercept form to write the equation of the line.

Slope-Intercept Form: $y = mx + b$

Example: What is the equation of the line in standard form that has a y-intercept of 2 and a slope of $-\frac{1}{3}$?

    Solution: Write the equation in slope-intercept form and then transform it to standard form.

    The slope of the line is $-\frac{1}{3}$, so $m = -\frac{1}{3}$. The y-intercept is 2 so $b = 2$.

$\quad\quad y = mx + b \quad\quad$ Slope-intercept form.

$\quad\quad y = -\frac{1}{3}x + 2 \quad$ Substitute values.

$\quad\quad 3y = -x + 6 \quad\quad$ Multiply both sides of the equation by 3.

$\quad x + 3y = 6 \quad\quad$ Add x to both sides of the equation.

    In standard form, the equation of the line is $x + 3y = 6$.

> If you know the slope and the y-intercept of a line, then substitute the slope for m and the y-intercept for b in the slope-intercept form $y = mx + b$.

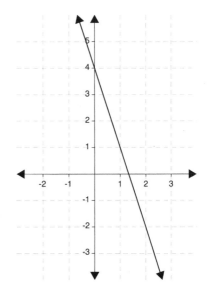

Practice Exercises

2.11   What are the slope and y-intercept of the line to the right?

2.12   What is the equation of the line in slope-intercept form? In standard form?

Point-Slope Form

If you know the slope of a line and a point on a line, or if you know two points on the line, you can use the *point-slope form* to find the equation of the line.

The point-slope form is derived from the slope formula. Recall that $Slope = \dfrac{(y_2 - y_1)}{(x_2 - x_1)} = m$. By multiplying both sides of the equation by

$(x_2 - x_1)$ we can rearrange the formula as $y_2 - y_1 = m(x_2 - x_1)$.

Point-slope form: $y - y_1 = m(x - x_1)$

Example: What is the equation of the line in standard form that passes through $(4, -6)$ with slope 3?

> **Solution:** Substitute the given information into the point-slope form. Transform the equation into standard form.
>
> $m = 3, (x_1, y_1) = (4, -6)$
>
> $y - y_1 = m(x - x_1)$ Point-slope form.
>
> $y - (-6) = 3(x - 4)$ Substitute values.
>
> $y + 6 = 3x - 12$ Simplify.
>
> $-3x + y = -18$ Subtract $3x$ and 6 from both sides.
>
> $3x - y = 18$ Multiply both sides by -1.
>
> The equation of the line that passes through $(4, -6)$ with a slope of 3 is $3x - y = 18$.

You can also use point-slope form to find the equation of a line using two points. First use the two points to find the slope of the line, and then use the slope and one of the points in the point-slope form just as in the previous example.

Example: What is the equation of the line in standard form that passes through $(-1, 4)$ and $(2, 3)$?

> **Solution:** Use the two points to find the slope. Substitute the slope and either point into the point-slope form. Transform the equation into standard form.

$$m = \frac{(y_2 - y_1)}{(x_2 - x_1)} \qquad \text{Slope formula.}$$

$$m = \frac{3-4}{2-(-1)} = \frac{-1}{3} = -\frac{1}{3} \qquad \text{Substitute values and simplify.}$$

$$m = -\frac{1}{3}, (x_1, y_1) = (-1, 4)$$

$$y - y_1 = m(x - x_1) \qquad \text{Point-slope form.}$$

$$y - 4 = -\frac{1}{3}(x - (-1)) \qquad \text{Substitute values.}$$

$$3y - 12 = -(x + 1) \qquad \text{Multiply both sides by 3.}$$

$$3y - 12 = -x - 1 \qquad \text{Use the Distributive Property.}$$

$$x + 3y = 11 \qquad \text{Add } x \text{ and 12 to both sides of the equation.}$$

The equation of the line that passes through $(-1, 4)$ and $(2, 3)$ is $x + 3y = 11$.

Practice Exercises

2.13 What is the equation of the line in standard form that passes through $(-2, 4)$ with slope 3?

2.14 What is the equation of the line in standard form that passes through $(2, -3)$ and $(1, 4)$?

The most common forms for a linear equation are slope-intercept form, point-slope form, and standard form.

Slope-Intercept Form: $y = mx + b$

• Use slope and y-intercept to graph a line.
• Use slope and y-intercept to write equation of a line.

Point-Slope Form: $y - y_1 = m(x - x_1)$

• Use point and slope to write equation of a line.
• If given two points, calculate slope and use slope and one point to write equation of a line.

Standard Form: $Ax + By = C$, where $A \geq 0$; A, B, and C are integers; and A and B are not both 0.

- Substitute $x = 0$ and then $y = 0$ to find the x- and y-intercepts to graph.
- Often the form of choice for the equation of a line.

Parallel and Perpendicular Lines

Parallel lines have the same slope. The product of the slopes of perpendicular lines is –1. Notice the slope of each line as given in the slope-intercept equation of each line in Figure 2.2.

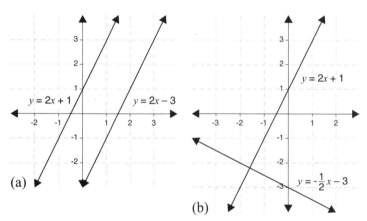

(a)

(b)

Figure 2.2
Slopes of parallel and perpendicular lines

The parallel lines in graph (a) are given by the equations $y = 2x + 1$ and $y = 2x - 3$. The slope of both lines is 2. Since the slopes are the same, the lines are parallel. This makes intuitive sense, because if two lines have the same steepness, they will never cross unless they are the same line.

The perpendicular lines in graph (b) are given by the equations $y = 2x + 1$ and $y = -\frac{1}{2}x - 3$. The slopes of the lines are 2 and $-\frac{1}{2}$, and $2 \cdot -\frac{1}{2} = -1$. Another way to think of the slopes of perpendicular lines is that they are reciprocals with opposite signs.

The slopes of parallel lines are equal.
The slopes of perpendicular lines are opposite reciprocals.

Example: What are the slopes of lines that are parallel to and perpendicular to $y = 4x + 1$?

Solution: Identify the slope of the given line. A parallel line will have the same slope. The product of the slope and the slope of a line perpendicular will be -1.

The slope of $y = 4x + 1$ is 4.

The slope of a line parallel is the same, or 4.

$4m = -1$

$m = -\dfrac{1}{4}$ Divide both sides by 4.

The slope of a line parallel to $y = 4x + 1$ is 4. The slope of a line perpendicular is $-\dfrac{1}{4}$.

You can use the relationship between the slopes of parallel and perpendicular lines to write equations of a line parallel or perpendicular to a given line.

Example: Write the equation of a line in standard form that is parallel to $y = 4x + 1$ and passes through the point $(1, -3)$.

Solution: Identify the slope of the line. A line parallel will have the same slope. Substitute the slope and the point into the point-slope form. Transform the equation to standard form.

The slope of $y = 4x + 1$ is 4. The slope of a line parallel is the same, or 4.

$y - y_1 = m(x - x_1)$ Point-slope form.

$y - (-3) = 4(x - 1)$ Substitute values.

$y + 3 = 4x - 4$ Simplify.

$-4x + y = -7$ Subtract $4x$ and 3 from both sides of the equation.

$4x - y = 7$ Multiply both sides of the equation by -1.

The equation of the line parallel to $y = 4x + 1$ and passing through $(1, -3)$ is $4x - y = 7$.

Practice Exercises

2.15   Write the equation of a line in standard form that is perpendicular to $y = 4x + 1$ and passes through the point $(1, -3)$.

2.16   Write the equation of a line in standard form that is parallel to $2x + 3y = -6$ and passes through $(-1, 4)$.

2.3 Direct Variation

When one variable in an equation is a constant multiple of the other, the two variables are said to be in direct proportion or *direct variation*. In a direct variation, the variables increase or decrease together. Furthermore, if the value of one variable is 0, so is the value of the other.

There are many real-world direct variations. The price of an item and the sales tax charged for the item, the diameter of a circle and its circumference, and the number of calories ingested from eating pepperoni pizza and the number of pieces of pepperoni pizza eaten are all examples of direct variation.

The direct variation relationship lends itself to an intuitive understanding of independent and dependent variables. The *dependent variable*, noted by y, is the quantity that depends on the other, the *independent variable*, noted by x. For example, the amount of calories ingested from eating pepperoni pizza depends on the number of pieces of pepperoni pizza eaten, so the number of calories is the dependent variable, and the number of pieces of pepperoni pizza is the independent variable.

A direct variation equation has the form $y = kx$, where k is the *constant of proportionality*. If you compare this equation to slope-intercept form, you will notice that the y-intercept of a direct variation is 0, and that k, the constant of proportionality is the same as m, the slope.

Slope-intercept form: $y = mx + b$

Direct variation form: $y = kx$

To find the equation for a direct variation, all you need to know is one point.

Example: The amount of sales tax charged on an item that costs $6 is
$0.33. What is the rate of sales tax? Write the direct variation
equation. Use the equation to find the amount of sales tax
charged on an item that costs $10.99.

Solution: The amount of sales tax depends on, or varies directly
with, the price of the item. Therefore, the price of the item is the
independent variable, and the sales tax is the dependent variable.
Substitute the cost of the item for x and the sales tax for y in the
direct variation equation. Solve for k. Write the direct variation
equation. Substitute $10.99 for x and find y.

$y = kx$	Direct variation equation.
$\$0.33 = k(\$6)$	Substitute values.
$0.055 = k$	Divide both sides by 6.

The sales tax rate is 0.055, or 5.5%.

$y = 0.055x$	Direct variation equation.
$y = 0.055(\$10.99)$	Substitute values.
$y = \$0.60445 \approx \$0.60$	Simplify.

The sales tax rate is 5.5%. The direct variation equation is
$y = 0.055x$. The sales tax on an item that costs $10.99 is $0.60.

Practice Exercises

2.17   The amount of simple interest on a loan varies directly with the
amount of the loan. If the annual interest on a loan of $1,000 is
$80, find the rate of interest, the direct variation equation, and
the amount of annual interest charged on a loan of $4,500 at the
same interest rate.

2.4 Scatter Plots and Lines of Regression

A *scatter plot* is a graph of points that shows the relationship between
two sets of data. The points on the scatter plot may show a general trend
in the data. The correlation between the data sets may show a *positive
correlation*, a *negative correlation,* or *no correlation*, as illustrated in
Figure 2.3.

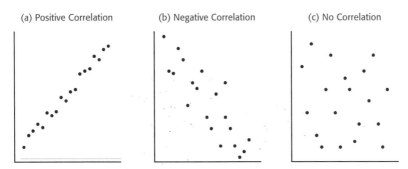

Figure 2.3 Correlation of data sets

The correlation of the data in a scatter plot can also be described as a weak or strong correlation. Graph (a) in Figure 2.3 shows a strong correlation, because the data points are spaced close together along a line. Graph (b) in Figure 2.3 shows a weaker correlation, because while the data shows a decreasing trend, the points are more widely spaced.

If a scatter plot shows a linear trend in the data, you can find a *line of best fit*. To estimate a line of best fit, draw a line through two data points that appears to represent the data well. A good line of best fit will lie as close as possible to all points and generally have about the same number of points above and below the line.

You can use the equation of a line of best fit to make predictions about the data.

Example: The following data shows the relationship between the duration of the previous eruption and the number of minutes until the next eruption of the Old Faithful geyser in Yellowstone National Park. Make a scatter plot of the data and draw a line of best fit through two data points. Use the coordinates of the data points to calculate the equation of the line of best fit. Use the equation of the line of best fit to predict the interval until the next eruption if the duration of the previous eruption is 3.7 minutes.

Duration (min.)	1.5	2.0	2.5	3.0	3.5	4.0	4.5	5.0
Interval (min.)	50	57	65	71	76	82	89	95

Solution: Plot the data points on a graph. Draw a line of best fit between two data points and find the equation of the line.

Substitute 3.7 for the appropriate variable and find the interval until the next eruption.

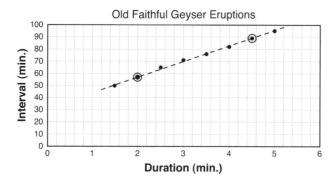

Line of best fit drawn through (2, 57) and (4.5, 89).

$$m = \frac{(y_2 - y_1)}{(x_2 - x_1)}$$ Slope formula.

$$m = \frac{(89 - 57)}{(4.5 - 2)} = \frac{32}{2.5} = 12.8$$ Substitute values.

$y - y_1 = m(x - x_1)$ Point-slope form.

$y - 57 = 12.8(x - 2)$ Substitute values.

$y - 57 = 12.8x - 25.6$ Use the Distributive Property.

$y = 12.8x + 31.4$ Add 57 to both sides.

The equation of the estimated line of best fit is $y = 12.8x + 31.4$ where x represents the duration of the previous eruption and y represents the interval until the next eruption.

If the previous eruption is 3.7 minutes, substitute 3.7 for x and find y.

$y = 12.8x + 31.4$

$y = 12.8(3.7) + 31.4$

$y = 78.76$

If the previous eruption time is 3.7 minutes, the interval until the next eruption will be about 79 minutes.

The equation of the estimated line of best fit will vary based on the two points chosen. A complex formula can be used to calculate the line of best fit, but it is beyond the scope of this book. However, many calculators are available that can calculate the line of best fit. Consult your calculator's manual for how to enter the data and find the line of best fit.

Example: Use a calculator to find the line of best fit for the Old Faithful geyser data in the previous example. Use the equation of the line of best fit to predict the interval until the next eruption if the duration of the previous eruption is 3.7 minutes.

Solution: Input the data and use the statistics function of the calculator to find the line of best fit. Substitute 3.7 for the appropriate variable and find the interval until the next eruption.

$y = 12.64x + 32.04$

$y = 12.64(3.7) + 32.04$

$y = 78.81$

If the previous eruption time is 3.7 minutes, the interval until the next eruption will be about 79 minutes.

When you use a calculator to find a line of best fit for data, it should provide the correlation coefficient, which is an indicator of whether the data shows a strong or weak correlation. A correlation coefficient of 1 or -1 indicates a perfect correlation, or that the data lines up in a perfectly straight line, while 0 indicates no correlation. In the previous example of the Old Faithful geyser, the correlation coefficient is 0.99685, which indicates a strong, positive correlation. This is supported by the graph, which shows data that is increasing and closely aligned to a straight line.

Any predictions made using the line of best fit are predictions, not absolute answers. Data can be *interpolated* to make predictions that lie inside the range of the data set, such as in the Old Faithful geyser examples. Data can be *extrapolated* to make predictions that lie outside the range of the data set. If you used the given data to calculate the time until the next eruption if the duration of the previous eruption was 10 minutes, you would be extrapolating the data. Extrapolated data can be suspect because you are assuming that the relationship between the data behaves consistently outside the bounds of the given data, which may or may not be true. Furthermore, some values that lie outside of the data given may not make sense in real life.

Practice Exercises

The data in the table shows the average number of chirps per second that crickets made at different temperatures.

Temp. (°F)	88.6	71.6	93.3	84.3	80.6	75.2	69.7	79.6
Average number of chirps	20	16	19.8	18.4	17.1	15.5	14.7	15

2.18 Make a scatter plot of the data and estimate a line of best fit. Find the equation for the estimated line of best fit.

2.19 If available, use a graphing calculator to calculate the line of best fit.

2.20 Use the equations for the estimated line of best fit and the calculated line of best fit to predict the number of cricket chirps to expect at 65 degrees Fahrenheit.

2.21 Use the equations for the estimated line of best fit and the calculated line of best fit to predict the temperature if you hear a cricket chirp 22 times in a second.

2.5 Graphing Linear Inequalities

Just as the solutions to a linear inequality in one variable can be graphed on a number line, the solutions of a linear inequality in two variables can be graphed on a coordinate plane.

Every linear inequality has a related linear equation.

$y < 3x + 2$ Inequality.

$y = 3x + 2$ Related equation.

To graph the solutions of an inequality, first graph the related equation to find the *boundary* of the region represented by the solutions. The boundary will divide the coordinate plane into two half-planes. Shade the half-plane that represents the solutions.

The boundary line will be solid if the inequality is ≤ or ≥ since the boundary line is part of the solution.

The boundary line will be dashed if the inequality is < or > since the boundary line is *not* part of the solution.

Example: Graph $y < 3x + 2$

> **Solution:** Graph the related equation using a dashed line. Choose two points, one on each side of the boundary line. Substitute each of the points in the inequality to determine whether the point is a solution. Shade the half-plane that contains the point that is a solution to the inequality.
>
> The related equation $y = 3x + 2$ has a y-intercept of 2 and a slope of 3.
>
> Check $(0, 0)$.   Check $(-1, 1)$.
>
> $0 < 3(0) + 2$ $1 < 3(-1) + 2$
>
> $0 < 2\ \checkmark$ $1 < -1$ FALSE
>
> $(0, 0)$ lies in the region that is a solution, so shade the half-plane that contains $(0, 0)$.

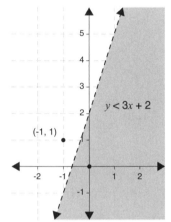

When you check points in the inequality, substitute the values into the original inequality. If the related equation was transformed using algebraic properties of equality, you cannot assume that the transformed equation will carry the same inequality symbol as the original inequality. Recall that if you multiply or divide both sides of an inequality by a negative number, the direction of the inequality changes.

Example: Graph $2x - 5y \geq 10$.

> **Solution:** Transform the related equation into slope-intercept form and graph the line using a solid line. Choose two points, one on each side of the boundary line. Substitute each of the points in the original inequality to determine whether the point is a solution. Shade the half-plane that contains the point that is a solution to the inequality.
>
> $2x - 5y = 10$ Related equation.
>
> $-5y = -2x + 10$ Subtract $2x$ from both sides.
>
> $y = \dfrac{2}{5}x - 2$ Divide both sides by -5.

Check (0, 0).   Check (2, –2).

$2(0) - 5(0) \geq 10$ $2(2) - 5(-2) \geq 10$

$0 \geq 10$ FALSE $14 \geq 10$ ✓

(2, –2) lies in the region that is a solution, so shade the half-plane that contains (2, –2).

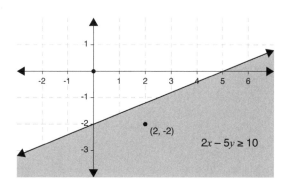

Practice Exercises

Graph the inequality.

2.22 $y > -2x + 5$

2.23 $3x + 4y \leq -8$

Chapter 3

Systems of Equations and Inequalities

3.1 Solving Systems of Equations

A *system of equations* is a set of two or more equations considered together. A solution of a system of equations must be a solution to all of the equations in the system.

System of Equations

$$\begin{cases} 3x + 2y = 5 \\ x - y = 0 \end{cases}$$

In the previous system of equations, $(1, 1)$ is a solution because $(1, 1)$ is a solution to both equations.

$$3x + 2y = 5 \qquad\qquad x - y = 0$$
$$3(1) + 2(1) = 5 \qquad\qquad 1 - 1 = 0$$

Solving a System of Equations by Graphing

One way to find the solution to a system of equations is to graph the equations. Recall that a graph is a visual representation of the solution to an equation. The solution to a system of equations is the point or points where the graphs of the equations intersect.

A system of linear equations will have one solution, no solution, or an infinite number of solutions.

One solution: The lines intersect at one point.

No solution: The lines are parallel.

Infinite number of solutions: The equations describe the same line.

Example: Solve the system of equations by graphing.

$$\begin{cases} x - 3y = 6 \\ 2x + 3y = 3 \end{cases}$$

Solution: Solve each equation for y and graph both equations on the same coordinate plane. Find the point of intersection.

$$x - 3y = 6 \qquad\qquad 2x + 3y = 3$$
$$-3y = -x + 6 \qquad\qquad 3y = -2x + 3$$
$$y = \frac{1}{3}x - 2 \qquad\qquad y = -\frac{2}{3}x + 1$$

$$\text{Slope} = \frac{1}{3} \qquad \text{Slope} = -\frac{2}{3}$$

y-intercept $= -2$ y-intercept $= 1$

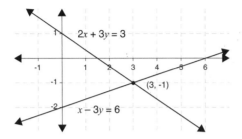

The point of intersection is $(3, -1)$.

Check:

$x - 3y = 6$	$2x + 3y = 3$
$3 - 3(-1) = 6$	$2(3) + 3(-1) = 3$
$6 = 6 \checkmark$	$3 = 3 \checkmark$

The solution to the system of equations is $(3, -1)$.

A system of equations with at least one solution is called a *consistent* system, while a system with no solutions is called an *inconsistent* system.

A consistent system with *exactly* one solution is called *independent*, while a consistent system with an infinite number of solutions is called *dependent*.

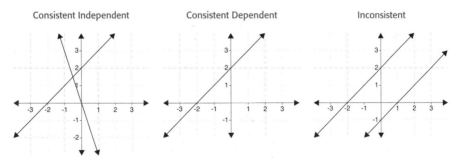

In the previous example, the system was consistent and independent because there was exactly one solution.

Example: Solve the system of equations by graphing and identify the system as consistent or inconsistent. If the system is consistent, identify it as independent or dependent.

$$\begin{cases} 2x + 2y = -6 \\ y = -x + 2 \end{cases}$$

Solution: Graph both equations on the same coordinate plane. Find the point(s) of intersection.

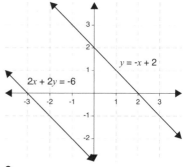

$2x + 2y = -6$

$ 2y = -2x - 6$

$y = -x - 3$	$y = -x + 2$
Slope $= -1$	Slope $= -1$
y-intercept $= -3$	y-intercept $= 2$

Notice that both lines have a slope of -1. Since they have different y-intercepts, the lines are parallel, and the system is inconsistent.

Practice Exercises

Solve by graphing. Identify the system as consistent or inconsistent. If the system is consistent, identify it as independent or dependent.

3.1 $\begin{cases} x + 2y = -4 \\ 3x - y = 2 \end{cases}$

3.2 $\begin{cases} 2x - y = -4 \\ -6x + 3y = 12 \end{cases}$

Solving a System of Equations by Substitution

Graphing a system of equations can be cumbersome and sometimes will provide only an approximation of the solution to the system. Using algebraic methods will lead to exact solutions. One algebraic method that works well when one of the coefficients is 1 or −1 is the substitution method.

To use the *substitution method,*

- Solve one equation for one variable.
- Substitute the expression that the variable is equal to into the other equation.
- Solve the new equation to find the value of one variable.

- Substitute the known value into one of the original equations.
- Solve for the other variable.

Example: Solve the system of equations by substitution. Identify the system as consistent or inconsistent. If the system is consistent, identify it as independent or dependent.

$$\begin{cases} x - 3y = 6 \\ 2x + 4y = 2 \end{cases}$$

Solution: Solve the first equation for x since it has a coefficient of 1.

$x - 3y = 6$

$\quad x = 3y + 6$

Substitute $3y + 6$ for x in the second equation and solve for y.

$\quad 2x + 4y = 2$

$2(3y + 6) + 4y = 2$

$\quad 6y + 12 + 4y = 2$

$\qquad 10y = -10$

$\qquad y = -1$

Substitute -1 for y in one of the original equations and solve for x.

$\quad x - 3y = 6$

$x - 3(-1) = 6$

$\quad x + 3 = 6$

$\qquad x = 3$

The solution to the system of equations is $(3, -1)$, and the system is consistent and independent since it has exactly one solution.

You can check the solution by substituting $(3, -1)$ in both equations.

Practice Exercises

Solve by substitution.

3.3 $\begin{cases} 3x + y = 7 \\ 2x + 3y = 7 \end{cases}$

3.4 $\begin{cases} 4x + 2y = 4 \\ 3x + y = 2 \end{cases}$

Solving a System of Equations by Elimination

If there is no variable in a system of equations that has a coefficient of 1, using the substitution method can lead to very complicated expressions with lots of fractions. In this case, the elimination method will often yield results faster and more efficiently.

To use the *elimination method,*

- Multiply one or both of the equations by a number that will create opposite coefficients for the same variable in the equations.

- Add the equations together so that one variable is eliminated.

- Solve for the remaining variable.

- Substitute the known value into one of the original equations.

- Solve for the other variable.

Example: Solve the system of equations by elimination.

$$\begin{cases} -2x + 3y = -5 & \text{Equation 1.} \\ 3x + 6y = 18 & \text{Equation 2.} \end{cases}$$

Solution: Multiply Equation 1 by –2 so that the coefficient of y is the opposite of the coefficient of y in Equation 2.

$$-2x + 3y = -5 \quad \longrightarrow \quad 4x - 6y = 10 \quad \text{Multiply Equation 1 by –2.}$$
$$\underline{3x + 6y = 18} \quad \text{Equation 2.}$$
$$7x = 28 \quad \text{Add equations.}$$
$$x = 4$$

$$3x + 6y = 18 \quad \text{Equation 2.}$$
$$3(4) + 6y = 18 \quad \text{Substitute value of } x.$$
$$12 + 6y = 18$$
$$6y = 6$$
$$y = 1$$

The solution to the system is (4, 1).

The check is left to you.

You may have to multiply both equations by a number in order to have the same variable in both equations that are opposites.

Example: Solve the system of equations by elimination.

$$\begin{cases} 2x + 3y = 1 & \text{Equation 1.} \\ 3x + 2y = 4 & \text{Equation 2.} \end{cases}$$

Solution: One way to solve the system is to multiply Equation 1 by 3 and Equation 2 by -2 so that the coefficients of x will be opposites.

$$2x + 3y = 1 \quad \longrightarrow \quad 6x + 9y = 3 \qquad \text{Multiply Equation 1 by 3.}$$

$$3x + 2y = 4 \quad \longrightarrow \quad \underline{-6x - 4y = -8} \qquad \text{Multiply Equation 2 by } -2.$$

$$5y = -5 \qquad \text{Add equations.}$$

$$y = -1$$

$$2x + 3y = 1 \qquad \text{Equation 1.}$$

$$2x + 3(-1) = 1 \qquad \text{Substitute value of } y.$$

$$2x - 3 = 1$$

$$2x = 4$$

$$x = 2$$

The solution to the system is $(2, -1)$.

The check is left to you.

Practice Exercises

Solve the system of equations by elimination.

3.5 $\quad \begin{cases} -3x + 4y = -5 \\ 5x - 2y = 27 \end{cases}$

3.6 $\quad \begin{cases} 2x + 6y = 26 \\ 7x - 4y = -34 \end{cases}$

Systems with No Solution or Infinite Solutions

If a system of linear equations has no solution or an infinite number of solutions, the substitution or elimination method will cancel out the variables, leaving either a true or false statement, such as $3 = 3$ or $0 = 1$.

If the statement is false, the system has no solution; the equations describe parallel lines, and the system is inconsistent.

If the statement is true, there are infinitely many solutions; the equations describe the same line, and the system is consistent and dependent.

Example: Solve the system of equations. Identify the system as consistent or inconsistent. If the system is consistent, identify it as independent or dependent.

$$\begin{cases} 2x - y = -4 \\ -6x + 3y = 12 \end{cases}$$

Solution: Since one equation has a variable with a coefficient of 1 or -1, substitution will be a simple way to solve. Solve the first equation for y and substitute the value of y into the second equation.

$2x - y = -4$

$-y = -2x - 4$

$y = 2x + 4$

Substitute $2x + 4$ for y in the second equation.

$-6x + 3y = 12$

$-6x + 3(2x + 4) = 12$

$-6x + 6x + 12 = 12$

$12 = 12$

The remaining statement is true. The equations are the same line with an infinite number of solutions, and the system is consistent and dependent.

Notice that the system in the previous example is the same system for Practice Exercise 3.2. When you graphed the system in that problem, you should have found that the lines both have slope = 2 and y-intercept = 4.

Example: Solve the system of equations. Identify the system as consistent or inconsistent. If the system is consistent, identify it as independent or dependent.

$$\begin{cases} 3x + 2y = 3 \\ -6x - 4y = 7 \end{cases}$$

Solution: Neither equation has a variable with a coefficient of 1 or −1. Elimination will be the simplest way to solve the system. Multiply the first equation by 2 and add the result to the second equation.

$$3x + 2y = 3 \quad \longrightarrow \quad 6x + 4y = 6 \quad \text{Multiply Equation 1 by 2.}$$

$$\underline{-6x - 4y = 7} \quad \text{Equation 2.}$$

$$0 = 13 \quad \text{Add equations.}$$

The statement is false, so the system has no solution. The system is inconsistent.

Practice Exercises

Solve the system of equations. Identify the system as consistent or inconsistent. If the system is consistent, identify it as independent or dependent.

3.7 $\begin{cases} 2x - y = 6 \\ -4x + 2y = 3 \end{cases}$

3.8 $\begin{cases} 4x - 3y = -6 \\ 5x - 2y = -11 \end{cases}$

3.9 $\begin{cases} x + 3y = 10 \\ -2x - 6y = -20 \end{cases}$

3.2 Solving Systems of Inequalities by Graphing

The solution to a system of inequalities is all of the ordered pairs that satisfy the inequalities in the system. Typically, the solution set will be represented by a region on a graph.

To find the solution to a system of linear inequalities:

• Graph each inequality, shading the solution.

• Identify the region that is shaded for all of the inequalities.

Example: Solve the system of inequalities.

$$\begin{cases} y < 3x - 2 \\ 2x + 5y \geq 25 \end{cases}$$

Solution: Graph each related equation using a solid line or a dashed line. Substitute ordered pairs into the original inequality to identify the region to be shaded.

$y = 3x - 2$

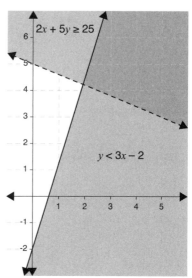

Slope = 3, y-intercept = -2

Check $(0, 0)$

$y < 3x - 2$

$0 < 3(0) - 2$

$0 < -2$ FALSE

Shade the region that does not include $(0, 0)$.

$2x + 5y = 25$

$\qquad 5y = -2x + 25$

$\qquad y = -\dfrac{2}{5}x + 5$

Slope = $-\dfrac{2}{5}$, y-intercept = 5

Check $(0, 0)$.

$\qquad 2x + 5y \geq 25$

$2(0) + 5(0) \geq 25$

$\qquad\qquad 0 \geq 25$ FALSE

Shade the region that does not include $(0, 0)$.

The solution to the system is the region represented by the overlap of the solutions to each inequality. $(5, 5)$ is in the solution set of the system.

Check $(5, 5)$.

$y < 3x - 2$	$2x + 5y \geq 25$
$5 < 3(5) - 2$	$2(5) + 5(5) \geq 25$
$5 < 13$ ✓	$35 \geq 25$ ✓

A system of two linear inequalities may have no solution. For a system of linear inequalities to have no solution, the related equations must be parallel lines.

Example: Solve the system of inequalities.

$$\begin{cases} y < x+1 \\ y > x+5 \end{cases}$$

Solution: Graph each related equation. Substitute ordered pairs to identify the region to be shaded.

$y = x + 1$

Slope $= 1$, y-intercept $= 1$

Check $(0, 0)$

$y < x + 1$

$0 < 0 + 1$

$0 < 1$ ✓

Shade the region that includes $(0, 0)$.

$y = x + 5$

Slope $= 1$, y-intercept $= 5$

Check $(0, 0)$

$y > x + 5$

$0 > 0 + 5$

$0 > 5$ FALSE

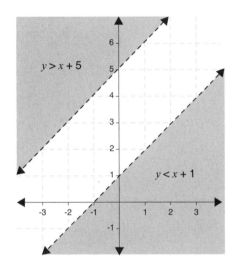

Shade the region that does not include $(0, 0)$.

The regions do not overlap, so the system has no solution.

Notice, however, in the previous example, that if the inequality symbols had been reversed, $y > x + 1$ and $y < x + 5$, the solution would have been the region between the two parallel lines.

A system of three linear inequalities may form a solution set that is a polygonal region. As the number of inequalities increases, it is easier to identify the solution region if you shade only the region that represents the solution to the system rather than all the regions that are solutions for each individual inequality.

Example: Solve the system of inequalities. Identify the coordinates of the vertices of the polygonal region that represents the solution.

$$\begin{cases} y \geq -2x + 2 \\ x \leq 1 \\ y \leq 3x + 1 \end{cases}$$

Solution: Graph each related equation. Substitute ordered pairs to identify the region to be shaded. Shade only the region that represents the final solution.

$y = -2x + 2$

Slope = -2, y-intercept = 2

Check (0, 0)

$y \geq -2x + 2$

$0 \geq -2(0) + 2$

$0 \geq 2$ FALSE

The solution region does not include (0, 0).

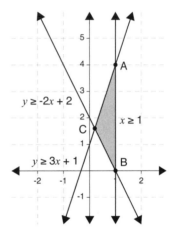

$x = 1$

Vertical line at $x = 1$.

Check (0, 0)

$x \leq 1$

$0 \leq 1$ ✓

The solution region includes (0, 0).

$y = 3x + 1$

Slope = 3, y-intercept = 1

Check (0, 0)

$y \leq 3x + 1$

$0 \leq 3(0) + 1$

$0 \leq 1$ ✓

The solution region includes (0, 0).

Vertices A and B are identifiable from the graph. Vertex C does not have integer coordinates. Use substitution or elimination to find the intersection of the two related equations.

$$y = -2x + 2$$

$$y = 3x + 1$$

$-2x + 2 = 3x + 1$ Substitute value of y from first equation into second.

$$1 = 5x$$

$$\frac{1}{5} = x$$

$$y = -2\left(\frac{1}{5}\right) + 2$$

$$y = \frac{8}{5}$$

$$A(1, 4),\ B(1, 0),\ C\left(\frac{1}{5}, \frac{8}{5}\right)$$

Practice Exercises

3.10 Solve the system of inequalities.

$$\begin{cases} y \le \frac{1}{2}x + 1 \\ y > -2x - 4 \end{cases}$$

3.11 Solve the system of inequalities.

$$\begin{cases} y < 3x + 1 \\ -3x + y > -4 \end{cases}$$

3.12 Solve the system of inequalities. Identify the coordinates of the vertices of the polygonal region that represents the solution.

$$\begin{cases} -2x + y > -3 \\ y < 5 \\ 4x + 3y > 3 \end{cases}$$

3.3 Linear Programming

Linear programming is a mathematical tool used in business to analyze and manage time and resources, often to maximize profit or minimize expenses. The particular situation is described by a set of inequalities, and the solution to the system of inequalities is called a *feasible region*. The equation that represents the profit or expenses is called the *objective equation*; typically, you are looking for maximum profit or minimum expenses. The maximum or minimum values will always occur at the vertices of the feasible region, so the coordinates of the vertices can be substituted into the objective equation to find the maximum or minimum values at the points at which they occur.

The objective equation may also be called the *objective function*. Refer to Chapter 4 for information on functions and function notation.

For real-world linear programming problems, the feasible region will typically be in the first quadrant.

To solve a linear programming problem:

• Write the inequalities that describe the scenario if they are not provided.

• Graph the system of inequalities and find the feasible region.

• Identify the coordinates of the vertices of the feasible region from the graph or by finding the points of intersection of the related equations.

• Substitute the coordinates of the vertices into the objective equation to find the coordinate that produces the minimum or maximum value.

Example: Graph the following inequalities and find the feasible region. Using the objective equation $E = 14x + 30y$, find the minimum and maximum values and the points at which they occur.

$x \geq 0,\ x \leq 5,\ y \geq 0,\ y \leq 8,\ 3x + 5y \geq 15$

Solution: Graph the related equation for each inequality and shade the feasible region. Identify the coordinates of the vertices of the feasible region and substitute the coordinates into the objective equation. Identify the minimum and maximum values for E and the points at which they occur.

$x = 0$; vertical line through $(0, 0)$.

$x = 5$; vertical line through $(5, 0)$.

$y = 0$; horizontal line through $(0, 0)$.

$y = 8$; horizontal line through $(0, 8)$.

$3x + 5y = 15$

$$5y = -3x + 15$$

$$y = -\frac{3}{5}x + 3; \text{ slope } = -\frac{3}{5},$$

y-intercept $= 3$

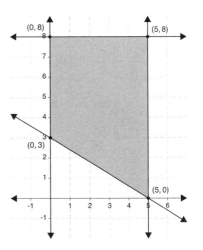

From the graph, the coordinates of the vertices are $(0, 3)$, $(0, 8)$, $(5, 8)$, and $(5, 0)$.

$E = 14(0) + 30(3) = 90$

$E = 14(0) + 30(8) = 240$

$E = 14(5) + 30(8) = 310$

$E = 14(5) + 30(0) = 70$

The minimum value is 70, given by the coordinates $(5, 0)$. The maximum value is 310, given by the coordinates $(5, 8)$.

You may have to consider a real-world problem and write the inequalities that describe the situation.

Example: Two carpenters and an assistant work together to make side tables and chairs to sell. It takes one carpenter 7 hours to make a side table and the assistant 1 hour to stain it. It takes one carpenter 2 hours to make a chair and the assistant 2 hours to stain it. The carpenters together work a maximum of 62 hours per week, and the assistant works a maximum of 26 hours per week on the side tables and chairs. Based on customer demand, the carpenters should make six or fewer chairs per side table. They earn $100 profit on every side table and $50 on every chair they make and sell. How many side tables and chairs should they make each week to maximize profit?

Solution: Write the inequalities and objective equation that fit the scenario. Graph the inequalities and find the feasible region. Substitute the coordinates of the vertices of the feasible region into the objective equation to find the maximum value.

Let x represent the number of tables and y represent the number of chairs.

$7x + 2y \leq 62$ 7 hr to make a table, 2 hr for chair, max. 62 hrs per week.

$x + 2y \leq 26$ 1 hr to finish table, 2 hr for chair, max. 26 hrs per week.

$y \leq 6x$ Make 6 or fewer chairs per table.

$y \geq 0$ Solution must be in first quadrant.

$x \geq 0$ Solution must be in first quadrant.

$P = 100x + 50y$ $100 profit per table, $50 per chair

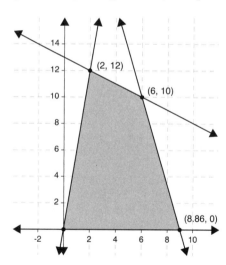

The vertex represented by the intersection of $y = 0$ and $7x + 2y = 62$ must be calculated. Substituting $y = 0$ into $7x + 2y = 62$ yields:

$7x + 2(0) = 62$

$\quad 7x = 62$

$\quad\quad x = \dfrac{62}{7} \approx 8.86$

$(8.86, 0)$

Vertices of the feasible region are (0, 0), (2, 12), (6, 10), and (8.86, 0).

Substitute the coordinates of each vertex into the objective equation.

$P = 100(0) + 50(0) = 0$

$P = 100(2) + 50(12) = 800$

$P = 100(6) + 50(10) = 1,100$

$P = 100(8.86) + 50(0) = 886$

The maximum profit of $1,100 is obtained by making and selling 6 side tables and 10 chairs.

Practice Exercises

3.13   Graph the following inequalities and find the feasible region. Using the objective equation $P = 4x + 18y$, find the minimum and maximum values and the points at which they occur.

$x + 2y \leq 10, x - 4y \leq -8, y \leq 3x - 2$

3.14   An artist is preparing two different handmade designs to sell. The first design takes 2 hours to make and earns a profit of $25. The second design takes 4 hours to make and earns a profit of $40. She believes she will sell at least twice as many of the second design as the first design. If she works at most 40 hours a week, how many of each design should she make to maximize her profit?

3.4 Systems of Equations in Three Variables

You have seen that an equation such as $2x + 3y = 2$ has two dimensions, x and y, and is represented by a line. An equation in three variables, such as $2x + 3y - z = 2$, has three dimensions, x, y, and z, and is represented by a plane in three-dimensional space.

Typically, a system of equations in three variables will have three equations, each representing a plane in space. The solution to a system of equations in three variables is represented by where all three planes intersect.

Graphing a system of equations in three variables is complicated and too cumbersome to use to find the solution. To solve a system of equations in three variables algebraically, you pair equations together, eliminate a variable, and create a system of two equations with only two variables. You can solve that system and then use substitution to find the value of the last variable.

To solve a system of equations in three variables:

• Pair two equations and use elimination to create an equation with only two variables.

• Pair two other equations and use elimination to create another equation with the same two variables as the first step.

• Solve the system of two equations in two variables using substitution or elimination.

• Substitute the value of the two known variables in one of the original equations to find the value of the last variable.

• Check the solution in all three original equations.

A system of equations in three variables can have one solution, no solution, or infinitely many solutions.

One Solution

A consistent, independent system has one solution, represented by the single point where the planes intersect.

Example: Solve the system of equations.

$$\begin{cases} 2x + 3y - z = 2 & \text{Equation 1.} \\ 3x - y + 2z = 4 & \text{Equation 2.} \\ x + 2y + 4z = 25 & \text{Equation 3.} \end{cases}$$

Solution: Pair two equations and use elimination to get rid of one of the variables. Use another set of equations and eliminate the same variable, creating a new system of two equations in two unknowns. Solve the system in two variables. Substitute the two known variables in one of the original equations to find the third variable. Check the solution.

For this solution, we will eliminate the variable z to create the system of two equations in two unknowns. You could choose to eliminate x or y instead.

Multiply Equation 1 by 2 and add to Equation 2 to eliminate z.

$2x + 3y - z = 2$ $\longrightarrow$ $4x + 6y - 2z = 4$ Multiply Equation 1 by 2.

$\underline{3x - y + 2z = 4}$ Equation 2.

$7x + 5y = 8$

Multiply Equation 1 by 4 and add to Equation 3 to eliminate z.

$2x + 3y - z = 2$ $\longrightarrow$ $8x + 12y - 4z = 8$ Multiply Equation 1 by 4.

$\underline{x + 2y + 4z = 25}$ Equation 3.

$9x + 14y = 33$

New system: $\begin{cases} 7x + 5y = 8 & \text{Equation 4.} \\ 9x + 14y = 33 & \text{Equation 5.} \end{cases}$

Multiply Equation 4 by 9 and Equation 5 by -7 to eliminate x.

$7x + 5y = 8$ $\longrightarrow$ $63x + 45y = 72$ Multiply Equation 4 by 9.

$9x + 14y = 33$ $\longrightarrow$ $\underline{-63x - 98y = -231}$ Multiply Equation 5 by -7.

$-53y = -159$

$y = 3$

Substitute y into one of the equations in the system of two equations to find x.

$7x + 5y = 8$ Equation 4.

$7x + 5(3) = 8$ Substitute the value of y.

$7x + 15 = 8$

$7x = -7$

$x = -1$

Substitute x and y into one of the original equations to find z.

$$2x + 3y - z = 2 \qquad \text{Equation 1.}$$
$$2(-1) + 3(3) - z = 2 \qquad \text{Substitute the value of } x \text{ and } y.$$
$$-2 + 9 - z = 2$$
$$7 - z = 2$$
$$-z = -5$$
$$z = 5$$

The solution to the system is $(-1, 3, 5)$. Check to verify that the solution satisfies all three original equations.

$$2x + 3y - z = 2 \qquad 3x - y + 2z = 4 \qquad x + 2y + 4z = 25$$
$$2(-1) + 3(3) - 5 = 2 \quad 3(-1) - 3 + 2(5) = 4 \quad -1 + 2(3) + 4(5) = 25$$
$$2 = 2 \checkmark \qquad\qquad 4 = 4 \checkmark \qquad\qquad 25 = 25 \checkmark$$

The solution to the system is $(-1, 3, 5)$.

No Solution or Infinitely Many Solutions

Just as you saw with systems of equations in two variables in Lesson 3.1, when you are solving a system of equations and the variables cancel out, you are working with a system that has either no solution or infinitely many solutions.

An inconsistent system has no solution, represented when all three planes do not intersect anywhere. Possibilities include when two or more planes are parallel or if two planes intersect and the third plane intersects the other two at different places.

Inconsistent System
No solution

A consistent, dependent system has an infinite number of solutions, represented when all three planes are the same plane or when the planes intersect in a line.

If at any point in solving a system of three equations, the variables are eliminated and you are left with a false statement, the system is inconsistent and has no solution.

Consistent, Dependent System
Infinite number of solutions

If the variables are eliminated and you are left with a true statement, the system will have infinitely many solutions with one exception. If two equations describe the same plane and the third plane is parallel, the system has no solution.

Example: Solve the system of equations.

$$\begin{cases} x + y + z = 4 & \text{Equation 1.} \\ 2x + 2y + 2z = 8 & \text{Equation 2.} \\ x + y + z = 6 & \text{Equation 3.} \end{cases}$$

Solution: For this solution, we will start by eliminating x.

Multiply Equation 1 by -2 and add it to Equation 2.

$x + y + z = 4 \quad \longrightarrow \quad -2x - 2y - 2z = -8 \quad$ Multiply Equation 1 by -2.

$\underline{2x + 2y + 2z = 8} \quad$ Equation 2.

$0 = 0$

Since the variables cancelled out and the statement is true, Equations 1 and 2 describe the same plane. Equation 3 may describe a plane that is parallel (no solution), or it may intersect the common plane (infinitely many solutions).

Multiply Equation 1 by -1 and add to Equation 3.

$x + y + z = 4 \quad \longrightarrow \quad -x - y - z = -4 \quad$ Multiply Equation 1 by -1.

$\underline{x + y + z = 6} \quad$ Equation 3.

$0 = 2$ FALSE

Since the variables cancelled out and the statement is false, Equation 3 describes a plane that is parallel to the plane described by Equations 1 and 2. The system has no solution.

Example: Solve the system of equations.

$$\begin{cases} 3x - 6y + 9z = -3 & \text{Equation 1.} \\ x - 2y + 3z = -1 & \text{Equation 2.} \\ -2x + 4y - 6z = 2 & \text{Equation 3.} \end{cases}$$

Solution: For this solution, we will start by eliminating x.

Multiply Equation 2 by 2 and add it to Equation 3.

$$x - 2y + 3z = -1 \quad \longrightarrow \quad 2x - 4y + 6z = -2 \quad \text{Multiply Equation 2 by 2.}$$
$$\underline{-2x + 4y - 6z = 2} \quad \text{Equation 3.}$$
$$0 = 0$$

Since the variables cancelled out and the statement is true, Equations 2 and 3 describe the same plane. Equation 1 may describe a plane that is parallel (no solution), or it may intersect the common plane (infinitely many solutions).

Multiply Equation 2 by –3 and add to Equation 1.

$$x - 2y + 3z = -1 \quad \longrightarrow \quad -3x + 6y - 9z = 3 \quad \text{Multiply Equation 2 by -3.}$$
$$\underline{3x - 6y + 9z = -3} \quad \text{Equation 1.}$$
$$0 = 0$$

Since the variables cancelled out and the statement is true, Equations 1 and 2 describe the same plane, and all three planes are the same plane.

The system has infinitely many solutions.

It is also possible to have infinitely many solutions when the three planes intersect in a line or when two planes are the same and that plane is intersected by the third plane.

> Remember: If at any time the variables cancel out and the statement is false, the system has *no solution*.
>
> If the variables cancel out and the statement is true, the system has *infinitely many solutions* unless two equations describe the same plane and the third plane is parallel.

Practice Exercises

Solve each system of equations.

3.15 $\begin{cases} 2x - y + 5z = 25 \\ -3x + 2y + z = -4 \\ 6x - 3y + 2z = 23 \end{cases}$

3.16 $\begin{cases} 2x - 4y + 6z = 5 \\ -x + 3y - 2z = -1 \\ x - 2y + 3z = 3 \end{cases}$

3.17 $\begin{cases} x + y + z = 0 \\ 4x + 4y + 2z = 5 \\ x + y - z = 5 \end{cases}$

3.5 Solving Systems of Equations with Matrices

A *matrix* is a rectangular array of numbers enclosed in brackets. Each number in the matrix is called an *element.* The plural of matrix is *matrices.*

3 × 3 matrix

$$\begin{bmatrix} 5 & 2 & -3 \\ 1 & 7 & -4 \\ -2 & 0 & 1 \end{bmatrix}$$

3 × 1 matrix

$$\begin{bmatrix} 1 \\ 6 \\ -3 \end{bmatrix}$$

You can write an *augmented matrix* to represent a system of equations in standard form by using the coefficients and the constants as elements in the matrix.

System of Equations $\longrightarrow$ Augmented Matrix

$$\begin{cases} 3x + y + z = 7 \\ 2x - 3y - 2z = -6 \\ x + 4y + z = 5 \end{cases} \longrightarrow \left[\begin{array}{ccc|c} 3 & 1 & 1 & 7 \\ 2 & -3 & -2 & -6 \\ 1 & 4 & 1 & 5 \end{array} \right]$$

A dashed or solid line separates the coefficients of the variables from the constants in each equation.

You can transform the augmented matrix using the following *row operations.*

- Interchange any two rows.

- Replace any row with a nonzero multiple of that row.

- Replace any row with the sum of that row and a nonzero multiple of another row.

To solve the system, write the augmented matrix and use row operations

to transform it to the form $\begin{bmatrix} 1 & d & e & | & g \\ 0 & 1 & f & | & h \\ 0 & 0 & 1 & | & j \end{bmatrix}$ where $d, e, f, g, h,$ and j are

real numbers. Then rewrite the augmented matrix as a system of equations and use substitution to finish solving.

Perform row operations to get the left entry of the top row to be a 1. Use that row to make the elements in the same column below to be 0.

Next, perform row operations to get the second element of the middle row to be a 1. Use that row to make the element below it 0.

Multiply by a constant to make the third element in the last row equal to 1.

You should always check your solution to make sure it works in all three equations.

Example: Solve the system $\begin{cases} x + 2y + z = 2 \\ 2x - y + 3z = -16 \\ x + 3y + 2z = 0 \end{cases}$

Solution: Write the augmented matrix and use row operations to simplify.

$\begin{bmatrix} 1 & 2 & 1 & | & 2 \\ 2 & -1 & 3 & | & -16 \\ 1 & 3 & 2 & | & 0 \end{bmatrix}$ Augmented matrix

$\begin{bmatrix} 1 & 2 & 1 & | & 2 \\ 0 & -5 & 1 & | & -20 \\ 1 & 3 & 2 & | & 0 \end{bmatrix}$ New Row 2 = -2(Row 1) + Row 2

$\begin{bmatrix} 1 & 2 & 1 & | & 2 \\ 0 & -5 & 1 & | & -20 \\ 0 & 1 & 1 & | & -2 \end{bmatrix}$ New Row 3 = -1(Row 1) + Row 3

$$\begin{bmatrix} 1 & 2 & 1 & | & 2 \\ 0 & 1 & 1 & | & -2 \\ 0 & -5 & 1 & | & -20 \end{bmatrix}$$ Exchange Row 2 and Row 3

$$\begin{bmatrix} 1 & 2 & 1 & | & 2 \\ 0 & 1 & 1 & | & -2 \\ 0 & 0 & 6 & | & -30 \end{bmatrix}$$ New Row 3 = 5(Row 2) + Row 3

$$\begin{bmatrix} 1 & 2 & 1 & | & 2 \\ 0 & 1 & 1 & | & -2 \\ 0 & 0 & 1 & | & -5 \end{bmatrix}$$ New Row 3 = $\frac{1}{6}$(Row 3)

Transform the augmented matrix back into a system of equations and use substitution to solve.

New system: $\begin{cases} x + 2y + z = 2 \\ \quad\quad y + z = -2 \\ \quad\quad\quad\quad z = -5 \end{cases}$

$z = -5$ $\quad\quad\quad$ $y + z = -2$ $\quad\quad\quad$ $x + 2y + z = 2$

$\quad\quad\quad\quad\quad$ $y + (-5) = -2$ $\quad$ $x + 2(3) + (-5) = 2$

$\quad\quad\quad\quad\quad\quad\quad$ $y = 3$ $\quad\quad\quad$ $x + 6 - 5 = 2$

$\quad\quad\quad\quad\quad\quad\quad\quad\quad\quad\quad\quad\quad\quad\quad\quad\quad$ $x = 1$

The solution to the system is $(1, 3, -5)$.

Check the values in the original system.

$x + 2y + z = 2$ $\quad\quad\quad$ $2x - y + 3z = -16$ $\quad\quad\quad$ $x + 3y + 2z = 0$

$1 + 2(3) + (-5) = 2$ $\quad$ $2(1) - 3 + 3(-5) = -16$ $\quad$ $1 + 3(3) + 2(-5) = 0$

$2 = 2$ ✓ $\quad\quad\quad\quad\quad$ $-16 = -16$ ✓ $\quad\quad\quad\quad\quad$ $0 = 0$ ✓

The solution to the system is $(1, 3, -5)$.

Practice Exercises

Solve each system of equations.

3.18 $\quad \begin{cases} 3x + y + z = 7 \\ 2x - 3y - 2z = -6 \\ x + 4y + z = 5 \end{cases}$

3.19
$$\begin{cases} 2x - y + 5z = 25 \\ -3x + 2y + z = -4 \\ 6x - 3y + 2z = 23 \end{cases}$$

3.6 Solving Systems of Equations with Determinants and Cramer's Rule

Every square matrix has a real number associated with it, called the *determinant*.

2×2 matrix   Determinant

$$\begin{bmatrix} a & c \\ b & d \end{bmatrix} \longrightarrow \quad \begin{vmatrix} a & c \\ b & d \end{vmatrix} = ad - bc$$

Notice that a matrix has brackets around it, and the determinant has straight lines. A matrix does not have a single value, while the value of the determinant is the difference between the products of the entries of the diagonals.

Example: Find the determinant of the matrix.

$$\begin{bmatrix} 1 & 5 \\ 3 & -4 \end{bmatrix}$$

Solution: Write the determinant. Find the products of the elements on the diagonals and subtract.

$$\begin{vmatrix} 1 & 5 \\ 3 & -4 \end{vmatrix} = 1(-4) - 3(5) = -4 - 15 = -19$$

You can use determinants to solve systems of linear equations. This method is called Cramer's Rule. *Cramer's Rule* expresses the solution for each variable in a system as the quotient of two determinants.

The solution to $\begin{cases} a_1 x + b_1 y = k_1 \\ a_2 x + b_2 y = k_2 \end{cases}$ is $x = \dfrac{D_x}{D}$ and $y = \dfrac{D_y}{D}$ where

$$D = \begin{vmatrix} a_1 & b_1 \\ a_2 & b_2 \end{vmatrix} \text{ and } D \neq 0, \ D_x = \begin{vmatrix} k_1 & b_1 \\ k_2 & b_2 \end{vmatrix} \text{ and } D_y = \begin{vmatrix} a_1 & k_1 \\ a_2 & k_2 \end{vmatrix}.$$

Find D first. If $D = 0$, then the system has no unique solution.

Example: Solve the system using Cramer's Rule.

$$\begin{cases} 2x + 3y = 1 \\ 3x + 2y = 4 \end{cases}$$

Solution: Write and find the value of the determinants D, D_x, D_y, and D_z. Use the values to find the value of x and y.

$$D = \begin{vmatrix} 2 & 3 \\ 3 & 2 \end{vmatrix} = 2(2) - 3(3) = -5 \quad D \neq 0, \text{ so there is a unique solution.}$$

$$D_x = \begin{vmatrix} 1 & 3 \\ 4 & 2 \end{vmatrix} = 1(2) - 4(3) = -10$$

$$D_y = \begin{vmatrix} 2 & 1 \\ 3 & 4 \end{vmatrix} = 2(4) - 3(1) = 5$$

$$x = \frac{D_x}{D} = \frac{-10}{-5} = 2; \quad y = \frac{D_y}{D} = \frac{5}{-5} = -1$$

The solution to the system is $(2, -1)$. This same system is solved by elimination in Lesson 3.1, so you can check the solution and compare different methods.

You can also use Cramer's Rule and determinants to solve a system of equations in three variables.

The determinant of a 3×3 matrix $\begin{bmatrix} a_1 & b_1 & c_1 \\ a_2 & b_2 & c_2 \\ a_3 & b_3 & c_3 \end{bmatrix}$ is defined as

$$\begin{vmatrix} a_1 & b_1 & c_1 \\ a_2 & b_2 & c_2 \\ a_3 & b_3 & c_3 \end{vmatrix} = a_1 b_2 c_3 + a_2 b_3 c_1 + a_3 b_1 c_2 - a_1 b_3 c_2 - a_2 b_1 c_3 - a_3 b_2 c_1.$$

Because this rule is long, complicated, and difficult to remember, you can also find the value of a determinant by using a method called *expansion by minors*. For a 3×3 determinant, the value is found by using three 2×2 determinants, called minors. The minor of an element is the 2×2 determinant formed with the elements remaining after the row and column of the element are eliminated.

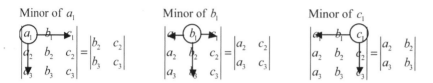

The elements of a single row or single column are multiplied by its minor and added or subtracted by the pattern at the right. If the sign in the position of the element is $+$, the product of the element and its minor are added. If the sign in the position of the element is $-$, the product of the element and its minor are subtracted.

$$\begin{vmatrix} + & - & + \\ - & + & - \\ + & - & + \end{vmatrix}$$

The value of the determinant expanding by minors about the first row looks like this:

$$\begin{vmatrix} a_1 & b_1 & c_1 \\ a_2 & b_2 & c_2 \\ a_3 & b_3 & c_3 \end{vmatrix} = a_1 \begin{vmatrix} b_2 & c_2 \\ b_3 & c_3 \end{vmatrix} - b_1 \begin{vmatrix} a_2 & c_2 \\ a_3 & c_3 \end{vmatrix} + c_1 \begin{vmatrix} a_2 & b_2 \\ a_3 & b_3 \end{vmatrix}.$$

The value of the determinant expanding by minors about the middle column looks like this:

$$\begin{vmatrix} a_1 & b_1 & c_1 \\ a_2 & b_2 & c_2 \\ a_3 & b_3 & c_3 \end{vmatrix} = -b_1 \begin{vmatrix} a_2 & c_2 \\ a_3 & c_3 \end{vmatrix} + b_2 \begin{vmatrix} a_1 & c_1 \\ a_3 & c_3 \end{vmatrix} - b_3 \begin{vmatrix} a_1 & c_1 \\ a_2 & c_2 \end{vmatrix}$$

No matter which row or column you expand by, the value of the determinant will be the same.

Example: Evaluate: $\begin{vmatrix} 1 & -1 & 4 \\ 3 & -2 & 0 \\ 1 & 1 & -3 \end{vmatrix}$. Evaluate by expanding about the

first row and by expanding about the second row.

Solution: Identify the minors and the product of the minor and each element. Simplify using the sign pattern.

Expanding about the first row:

$$\begin{vmatrix} 1 & -1 & 4 \\ 3 & -2 & 0 \\ 1 & 1 & -3 \end{vmatrix} = 1\begin{vmatrix} -2 & 0 \\ 1 & -3 \end{vmatrix} - (-1)\begin{vmatrix} 3 & 0 \\ 1 & -3 \end{vmatrix} + 4\begin{vmatrix} 3 & -2 \\ 1 & 1 \end{vmatrix}$$

$$= 1[(-2)(-3) - (1)(0)] + 1[(3)(-3) - (1)(0)] + 4[(3)(1) - (1)(-2)]$$
$$= 6 - 9 + 20$$
$$= 17$$

Expanding about the second row:

$$\begin{vmatrix} 1 & -1 & 4 \\ 3 & -2 & 0 \\ 1 & 1 & -3 \end{vmatrix} = -3\begin{vmatrix} -1 & 4 \\ 1 & -3 \end{vmatrix} + (-2)\begin{vmatrix} 1 & 4 \\ 1 & -3 \end{vmatrix} - 0\begin{vmatrix} 1 & -1 \\ 1 & 1 \end{vmatrix}$$

$$= -3[(-1)(-3) - (1)(4)] - 2[(1)(-3) - 1(4)] - 0$$
$$= 3 + 14$$
$$= 17$$

The value of the determinant is 17.

If you choose a row or column that has a zero to expand about, it will be easier to simplify. See the solution that expanded about the second row in the previous example.

Using Cramer's Rule, the solution of $\begin{cases} a_1x+b_1y+c_1z = d_1 \\ a_2x+b_2y+c_2z = d_2 \\ a_3x+b_3y+c_3z = d_3 \end{cases}$ is $x = \dfrac{D_x}{D}$,

$y = \dfrac{D_y}{D}$, $z = \dfrac{D_z}{D}$ where $D = \begin{vmatrix} a_1 & b_1 & c_1 \\ a_2 & b_2 & c_2 \\ a_3 & b_3 & c_3 \end{vmatrix}$, $D_x = \begin{vmatrix} d_1 & b_1 & c_1 \\ d_2 & b_2 & c_2 \\ d_3 & b_3 & c_3 \end{vmatrix}$,

$D_y = \begin{vmatrix} a_1 & d_1 & c_1 \\ a_2 & d_2 & c_2 \\ a_3 & d_3 & c_3 \end{vmatrix}$, $D_z = \begin{vmatrix} a_1 & b_1 & d_1 \\ a_2 & b_2 & d_2 \\ a_3 & b_3 & d_3 \end{vmatrix}$ and $D \neq 0$.

Example: Solve the system using Cramer's Rule. $\begin{cases} x+y+2z = -7 \\ 2x-3y+z = -2 \\ 5x+2y+3z = -7 \end{cases}$

Solution: Write and find the value of the determinants D, D_x, and D_y. Use the values to find the value of x, y, and z. Check the values in the system.

Expand each determinant about the first row.

$$D = \begin{vmatrix} 1 & 1 & 2 \\ 2 & -3 & 1 \\ 5 & 2 & 3 \end{vmatrix} = 1\begin{vmatrix} -3 & 1 \\ 2 & 3 \end{vmatrix} - 1\begin{vmatrix} 2 & 1 \\ 5 & 3 \end{vmatrix} + 2\begin{vmatrix} 2 & -3 \\ 5 & 2 \end{vmatrix} = 26$$

$$D_x = \begin{vmatrix} -7 & 1 & 2 \\ -2 & -3 & 1 \\ -7 & 2 & 3 \end{vmatrix} = -7\begin{vmatrix} -3 & 1 \\ 2 & 3 \end{vmatrix} - 1\begin{vmatrix} -2 & 1 \\ -7 & 3 \end{vmatrix} + 2\begin{vmatrix} -2 & -3 \\ -7 & 2 \end{vmatrix} = 26$$

$$D_y = \begin{vmatrix} 1 & -7 & 2 \\ 2 & -2 & 1 \\ 5 & -7 & 3 \end{vmatrix} = 1\begin{vmatrix} -2 & 1 \\ -7 & 3 \end{vmatrix} - (-7)\begin{vmatrix} 2 & 1 \\ 5 & 3 \end{vmatrix} + 2\begin{vmatrix} 2 & -2 \\ 5 & -7 \end{vmatrix} = 0$$

$$D_z = \begin{vmatrix} 1 & 1 & -7 \\ 2 & -3 & -2 \\ 5 & 2 & -7 \end{vmatrix} = 1\begin{vmatrix} -3 & -2 \\ 2 & -7 \end{vmatrix} - 1\begin{vmatrix} 2 & -2 \\ 5 & -7 \end{vmatrix} + (-7)\begin{vmatrix} 2 & -3 \\ 5 & 2 \end{vmatrix} = -104$$

$$x = \frac{D_x}{D} = \frac{26}{26} = 1 \; ; \; y = \frac{D_y}{D} = \frac{0}{26} = 0 \; ; \; z = \frac{D_z}{D} = \frac{-104}{26} = -4$$

The solution to the system is (1, 0, –4).

Check

$$x + y + 2z = -7 \qquad 2x - 3y + z = -2 \qquad 5x + 2y + 3z = -7$$

$$1 + 0 + 2(-4) = -7 \quad 2(1) - 3(0) + (-4) = -2 \quad 5(1) + 2(0) + 3(-4) = -7$$

$$-7 = -7 \checkmark \qquad\qquad -2 = -2 \checkmark \qquad\qquad -7 = -7 \checkmark$$

The solution to the system is (1, 0, –4).

Practice Exercises

Solve each system using Cramer's Rule.

3.20 $\begin{cases} 2x + 3y = -1 \\ x - 4y = 16 \end{cases}$

3.21 $\begin{cases} 2x - y + 5z = 25 \\ -3x + 2y + z = -4 \\ 6x - 3y + 2z = 23 \end{cases}$

Chapter 4

Relations and Functions

4.1 Definitions

In the first three chapters, you solved and graphed equations. Those equations, along with any set of ordered pairs, can be classified as relations. Some relations are also functions.

Relation

A *relation* is a set of input and output values. Relations might be shown as a set of ordered pairs, a mapping diagram, a graph, or an equation. Figure 4.1 shows five examples of relations. Relation (b) is a mapping diagram that shows the inputs in the left column and the outputs in the right column, and represents the ordered pairs (–3, 0), (2, 1), (4, 2) and (5, 1).

A relation is a *discrete relation* if the domain is a finite set of individual points, such as relation (a) in Figure 4.1. The graph of a discrete relation will be distinct points not connected by a line or curve. A relation is a *continuous relation* if the number of elements in the domain is infinite, such as relation (d) in Figure 4.1. The graph of a continuous relation will be a line or curve.

Example: Which relations in Figure 4.1 are discrete relations? Which relations in Figure 4.1 are continuous relations?

Solution: Relations (a), (b), and (c) are discrete relations because they are finite sets of distinct points. Relation (d) is a continuous relation because the graph is an infinite number of points connected by a line or curve. Relation (e) is continuous because any real number can be substituted for x to find y.

Domain and Range

The *domain* of a relation is the set of the first coordinates of the ordered pairs. The *range* of a relation is the set of the second coordinates of the ordered pairs. Typically the domain and range values of a discrete relation are listed in numerical order.

Example: Find the domain and range of relation (a) in Figure 4.1.

Solution: Domain = {1, 2, 3, 4}, Range = {–1, 1, 5, 6}

On a graph, the domain is represented by the values on the horizontal x-axis. The range is represented by the values on the vertical y-axis.

(a) [(1, 5), (2, −1), (3, 6), (4, 1)] (b)

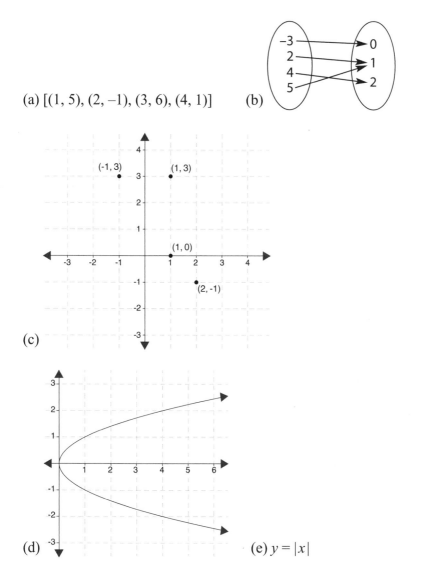

(c)

(d) (e) $y = |x|$

Figure 4.1 Examples of relations

Example: Find the domain and range of relation (c) in Figure 4.1.

Solution: Domain = {−1, 1, 2}, Range = {−1, 0, 3}

When a relation is described by an equation, the domain is the possible values of x, while the range is the possible values for y.

Example: Find the domain and range of relation (e) in Figure 4.1.

> **Solution:** In the equation $y = |x|$, you can substitute any real number for x. However, the only possible values for y are real numbers greater than or equal to 0 because the absolute value of any number will always be greater than or equal to 0.
>
> Domain = {all real numbers}, Range = {real numbers ≥ to 0}

Practice Exercises

4.1   Find the domain and range of relation (b) in Figure 4.1.

4.2   Find the domain and range of relation (d) in Figure 4.1.

Functions

A *function* is a relation in which each element in the domain is paired with one and only one element in the range. Another way to think of a function is that none of the domain, or x values, are repeated.

Example: Is relation (a) in Figure 4.1 a function?

> **Solution:** All of the x values, or the values in the domain, are different. Therefore, the relation is a function.

On a graph, you can use the *vertical line test* to quickly determine whether a relation is a function. If it is possible to pass a vertical line through two or more points on the graph, then the relation is *not* a function.

Example: Is relation (d) in Figure 4.1 a function? Explain.

> **Solution:** You can pass a vertical line through two points on the graph. The relation is not a function.

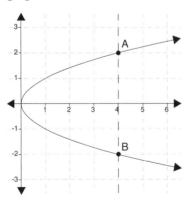

Practice Exercises

4.3   Is relation (b) in Figure 4.1 a function? Explain.

4.4   Is relation (c) in Figure 4.1 a function? Explain.

Equations can also be used to represent relations and functions. Recall from Chapter 2 that if an equation uses the variables x and y, then the solutions to the equation are the ordered pairs (x, y) that make the equation true.

Graphing the equation is often the best way to determine whether an equation represents a function.

Example: Is the relation $y = -\dfrac{2}{3}x + 1$ a function? Is the relation discrete or continuous?

Solution: Recall from Chapter 2 the slope-intercept form of a linear equation where $y = mx + b$, and m is the slope of the line, and b is the y-intercept. In $y = -\dfrac{2}{3}x + 1$, $m = -\dfrac{2}{3}$, and $b = 1$.

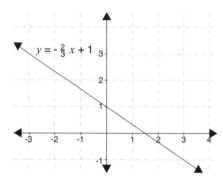

A vertical line passed through any part of the graph would touch only one point on the graph, so the relation is a function.

The graph is a line, so the relation is continuous.

Any linear relation, except a vertical line, is a function.

Recall that you can find solutions to an equation by choosing a value for one variable and solving for the other variable.

Example: Is the relation $y = x^2$ a function? Is the relation discrete or continuous?

> **Solution:** Choose several different values for x and substitute each value into the equation and find the value of y. Graph the ordered pairs and connect them with a line or smooth curve.

x	x^2	y
-2	$(-2)^2$	4
-1	$(-1)^2$	1
0	$(0)^2$	0
1	$(1)^2$	1
2	$(2)^2$	4

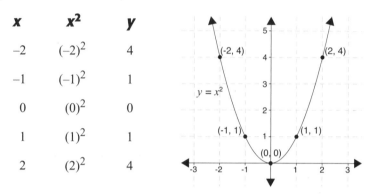

> A vertical line passed through any part of the graph would touch only one point on the graph, so the relation is a function.

> The graph is a curve, so it is continuous.

Practice Exercises

Is the relation a function? Is the relation discrete or continuous?

4.5 $y = 3x + 1$

4.6 $x = y^2$ (Hint: Substitute values for y to find the corresponding value for x.)

4.2 Function Notation

When an equation is a function, you can write the equation in *function notation*.

> $y = 3x + 1$ is written $f(x) = 3x + 1$ in function notation.

> $f(x) = 3x + 1$ is read as "f of x equals $3x + 1$."

> $f(x)$ means the value of the function at x.

> $f(x)$ is another name for y in an equation.

> Ordered pairs (x, y) can be written as $(x, f(x))$.

To find the value of the function at x, substitute a value for x and simplify.

Example: Find $f(-2)$ for $f(x) = 3x + 1$

    Solution: Substitute -2 for x and simplify.

$$f(x) = 3x + 1$$
$$f(-2) = 3(-2) + 1$$
$$f(-2) = -5$$

Make sure you use the order of operations when you evaluate functions!

You can also evaluate functions for other values.

Example: Find $f(x + h)$ for $f(x) = 2x - 4$.

    Solution: Substitute $(x + h)$ for x and simplify.

$$f(x) = 2x - 4$$
$$f(x + h) = 2(x + h) - 4$$
$$f(x + h) = 2x + 2h - 4$$

Example: Find $f(2b)$ for $f(x) = 3x^2$.

    Solution: Substitute $2b$ for x and simplify.

$$f(x) = 3x^2$$
$$f(2b) = 3(2b)^2$$
$$f(2b) = 3(4b^2)$$
$$f(2b) = 12b^2$$

In the notation $f(2)$, $f(x + h)$ or $f(2b)$, the value in the parentheses is a reminder of what you substituted in for the variable x in $f(x)$.

Practice Exercises

4.7   Evaluate $f(0)$ for $f(x) = x^2 + 2x - 3$

4.8   Find $f(x - h)$ for $f(x) = -x + 3$

4.9   Find $f(4y)$ for $f(x) = 6x + 2$

Equations given in function notation can be graphed. Remember that $f(x) = y$.

Example: Graph $f(x) = 2x - 5$.

> **Solution:** Since $f(x) = y$, $f(x) = 2x - 5$ is equivalent to $y = 2x - 5$. The equation is represented by a line. Using the information from the slope-intercept form $y = mx + b$, the slope of the line is 2, and the y-intercept is -5.

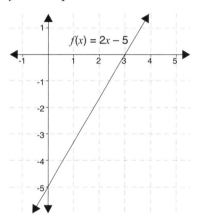

Example: Graph $f(x) = 2x^2 + 1$.

> **Solution:** Since the variable is raised to the second power, the equation is not represented by a line. Substitute values for x and calculate $f(x)$.

$f(x) = 2x^2 + 1$

x	$2x^2 + 1$	$f(x)$
-2	$2(-2)^2 + 1$	9
-1	$2(-1)^2 + 1$	3
0	$2(0)^2 + 1$	1
1	$2(1)^2 + 1$	3
2	$2(2)^2 + 1$	9

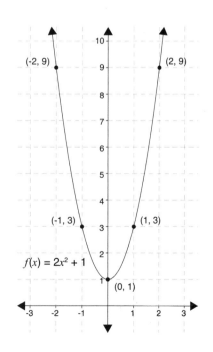

Practice Exercises

Graph each function.

4.10 $f(x) = \dfrac{2}{5}x + 3$

4.11 $f(x) = -x^2 - 4$

4.3 Piecewise Functions

A *piecewise function* is a function in which the definition of the function changes depending on the value of the independent variable. Several graphs of piecewise functions are shown in Figure 4.2.

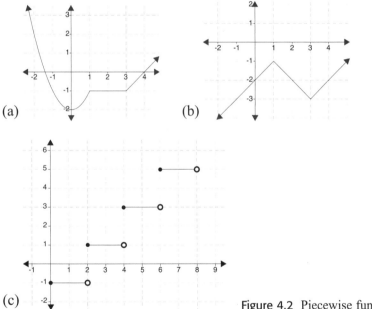

(a)

(b)

(c)

Figure 4.2 Piecewise functions

A *piecewise linear function* is a piecewise function whose pieces are linear, as illustrated by graphs (b) and (c) in Figure 4.2. A *step function* is a piecewise linear function made up of horizontal line segments that look like steps. Graph (c) in Figure 4.2 is a step function.

Notice the open and closed dots on graph (c). An open dot means that the point is not included in the solution, while a closed dot is a point that is included in the solution set. You will see open or closed dots when there is a break in the graph like you see in the step function graph.

Example: Graph the piecewise function and identify the domain and
   range.

$$f(x) = \begin{cases} -x^2 \text{ if } x \le 1 \\ x+2 \text{ if } x > 1 \end{cases}$$

Solution: Graph each piece of the function over the given interval
of x values. Where the graph experiences a break, determine
whether the endpoint should have an open or closed dot. Identify
the values included in the domain and the range.

$f(x) = -x^2$ if $x \le 1$ $f(x) = x + 2$ if $x > 1$

x	$-x^2$	$f(x)$
1	$-(1)^2$	-1
0	$-(0)^2$	0
-1	$-(-1)^2$	-1
-2	$-(-2)^2$	-4

x	$x + 2$	$f(x)$
1	$1 + 2$	3
2	$2 + 2$	4
3	$3 + 2$	5
4	$4 + 2$	6

$x = 1$ is included in the first piece of the function but not in the
second. Therefore, $(1, -1)$ will have a closed dot, and $(1, 3)$ will
have an open dot.

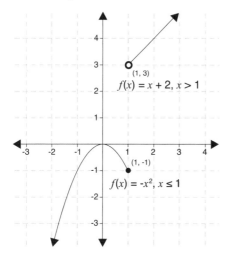

Recall that the domain is the set of all possible values for x. All real
numbers can be substituted for x, so the domain of the function is
all real numbers.

Recall that the range is the set of all possible values for $f(x)$. Notice the break in the graph. The possible values for $f(x)$ are real numbers greater than 3 or less than or equal to 0. Therefore, the range is $f(x) > 3$ or $f(x) \leq 0$.

Absolute Value Functions

The absolute value function is an example of a piecewise linear function because each piece of the function is linear. Recall from Chapter 1 that

$$|x| = \begin{cases} x \text{ if } x \geq 0 \\ -x \text{ if } x < 0 \end{cases}.$$

Therefore, the function $f(x) = |x| = \begin{cases} x \text{ if } x \geq 0 \\ -x \text{ if } x < 0 \end{cases}.$

Absolute value functions have a V-shaped graph.

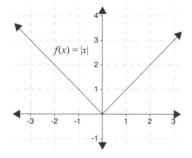

$f(x) = |x|$

Notice that the domain of the absolute value function is all real numbers, and the range of the absolute value function is all real numbers greater than or equal to 0.

Example: Graph the piecewise function and identify the domain and range.

$$f(x) = 2\,|x + 1|$$

Solution: Create a table of values using both positive and negative numbers. Graph the points and connect them.

x	2|x + 1|	f(x)
–3	2|–3 + 1|	4
–2	2|–2 + 1|	2
–1	2|–1 + 1|	0
0	2|0 + 1|	2
1	2|1 + 1|	4

The domain is all real numbers. The range is $f(x) \geq 0$.

Step Functions

Step functions are very common in pricing situations. For example, an auto repair shop may charge $45 per hour for a mechanic's time, including any part of an hour worked. So, if the mechanic works one and a half hours, you will be charged for 2 hours.

Example: The table below shows postage rates for letters based on the weight of the letter. Graph the step function.

Weight Not Over (oz.)	1	2	3	3.5
Postage		$0.46	$0.66	$0.86	$1.06

Solution: Identify the interval of weight for each postage rate. Graph the intervals.

x (weight)	$f(x)$ (postage)
$0 < x \le 1$	0.46
$1 < x \le 2$	0.66
$2 < x \le 3$	0.86
$3 < x \le 3.5$	1.06

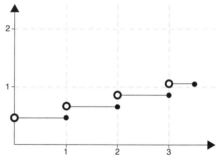

Practice Exercises

Graph the function and identify the domain and range.

4.12 $f(x) = \begin{cases} x - 3 \text{ if } x \le -1 \\ x + 2 \text{ if } x > -1 \end{cases}$

4.13 $f(x) = |x - 2|$

Graph the function.

4.14   An auto shop has a labor charge of $45 per hour or any part of an hour. Make a graph to show the labor charges for any job up to 5 hours of labor.

4.4 Parent Functions and Transformations

A useful way to organize your thinking about functions is to consider function families. A *function family* is a group of similar functions. Linear, quadratic, and absolute value functions are examples of three different function families. A *parent function* is the simplest function of its type. If you know the form of the parent function and how different values affect, or transform, the parent function, you can easily sketch a graph of the function.

Linear Functions

You have already seen the principle of transforming a parent function with a linear function. The parent function of a linear function is $f(x) = x$. (Remember that this is the same as $y = x$.) You learned in Lesson 2.1 that in $y = mx + b$, the slope, m, affects the steepness and direction of the line, while b is the y-intercept. Changing the value of b moves the line up or down the y-axis.

Figure 4.3 illustrates how changing the values of m and b transforms the line from the parent function.

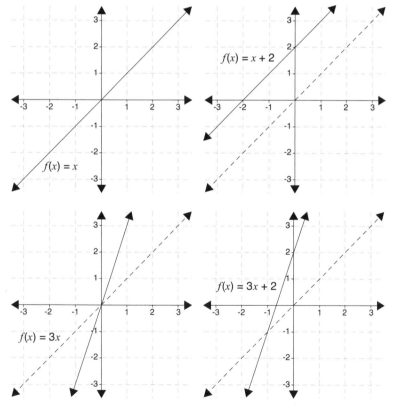

Figure 4.3 Transforming the linear parent function $f(x) = x$

Absolute Value and Quadratic Functions

The parent function of absolute value functions is $f(x) = |x|$. In Lesson 4.3, you saw that the graph of an absolute value function is V-shaped.

A quadratic function is a function in which the highest power of the independent variable is 2. The parent function of quadratic functions is $f(x) = x^2$.

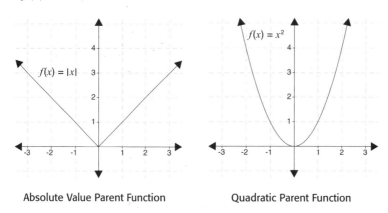

Absolute Value Parent Function   Quadratic Parent Function

Figure 4.4 Absolute value and quadratic parent functions

A *translation* slides a figure up, down, left, or right while keeping the same shape.

- A constant value k that is added to a parent function moves the figure up if k is positive, or down if k is negative. The function will have the form $y = f(x) + k$, such as $y = x^2 + 1$ or $y = |x| - 4$. Notice that k is "outside" the parent function.

- A constant value h that is *subtracted* from x before the parent function is evaluated moves the figure left if h is negative, and right if h is positive. The function will have the form $y = f(x - h)$, such as $y = (x + 1)^2$ or $y = |x - 4|$. Notice that h is "inside" the parent function.

Example: Sketch the graph of each function: $y = x^2 - 3$ and $y = |x + 2|$.

  Solution: $y = x^2 - 3$ is of the form $y = f(x) + k$ where $k = -3$. The parent function $f(x) = x^2$ will be translated 3 units down.

  $y = |x + 2|$ is of the form $y = f(x - h)$ where $h = -2$. The parent function $f(x) = |x|$ will be translated 2 units to the left.

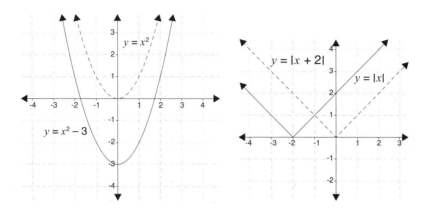

Practice Exercises

Use transformations to sketch the graph.

4.15 $y = |x| - 1$

4.16 $y = (x+4)^2$

A *reflection* flips the figure over a line of reflection while keeping the same shape.

- When a parent function is multiplied by -1, the figure is reflected over the x-axis. The function will have the form $y = -f(x)$, such as $y = -x^2$ or $y = -|x|$. Notice that the negative is "outside" the parent function.

Example: Sketch the graph of each function: $y = -|x|$ and $y = -(x-4)^2$

   Solution: $y = -|x|$ is of the form $y = -f(x)$. The parent function $f(x) = |x|$ will be reflected over the x-axis.

   $y = -(x-4)^2$ is of the form $y = -f(x-h)$. The parent function $f(x) = x^2$ will be translated 4 units to the right and reflected over the x-axis.

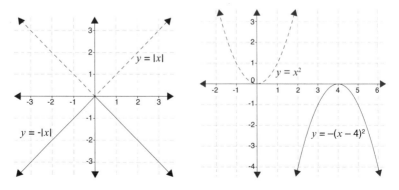

Practice Exercises

Use transformations to sketch the graph.

4.17 $y = -(x + 2)^2$

4.18 $y = -|x + 3|$

A *dilation* stretches or compresses a figure proportionally.

- When a parent function is multiplied by a constant a, the figure is dilated. The function will have the form $y = af(x)$, such as $y = 3x^2$ or $y = \frac{1}{2}|x|$. Notice that a is "outside" the parent function. If $|a| > 1$, the figure is stretched vertically and the figure looks "skinnier." If $0 < |a| < 1$, the figure is compressed vertically and the figure looks "fatter."

Example: Sketch the graph of each function: $y = 2|x|$ and $y = -\frac{1}{2}x^2$.

  Solution: $y = 2|x|$ is of the form $y = af(x)$ where $|a| > 1$. Since $a > 0$, there is no reflection. The parent function $f(x) = |x|$ will be stretched vertically and appear "skinnier."

  $y = -\frac{1}{2}x^2$ is of the form $y = af(x)$ where $a < 0$ and $0 < |a| < 1$. The parent function $f(x) = x^2$ will be reflected over the x-axis and compressed vertically and appear "fatter."

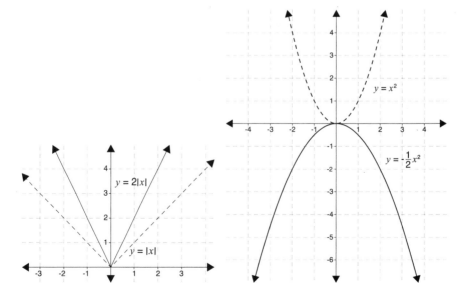

Practice Exercises

Use transformations to sketch the graph.

4.19 $y = -3x^2$

4.20 $y = \dfrac{1}{3}|x|$

The table below summarizes the transformations to a nonlinear function.

Parent Graph Transformations

Transformation	Sample Function	Parent Graph Change
Translation		
$f(x) + k, k > 0$	$y = x^2 + 1$	figure moves up k units
$f(x) + k, k < 0$	$y = |x| - 4$	figure moves down k units
$f(x - h), h < 0$	$y = (x + 1)^2$	figure moves left h units
$f(x - h), h > 0$	$y = |x - 4|$	figure moves right h units
Reflection		
$af(x), a < 0$	$y = -x^2$	figure is reflected over the x-axis
Dilation		
$af(x), 0 < |a| < 1$	$y = \dfrac{1}{2}|x|$	figure is compressed vertically
$af(x), |a| > 1$	$y = 3x^2$	figure is stretched vertically

A function with multiple transformations has the form $y = a(x - h)^2 + k$ or $y = a|x - h|^2 + k$. To graph a function with multiple transformations, perform the transformations to the graph in the same way you would apply the order of operations in the equation:

• Translation left or right as given by h.

• Reflection and/or dilation as given by a.

• Translation up or down as given by k.

Example: Sketch the graph of $y = -2(x-3)^2 + 4$

> **Solution:** The function is a quadratic function. The parent function $y = x^2$ is translated 3 units to the right. It is reflected over the x-axis and stretched vertically, and then translated 4 units up.

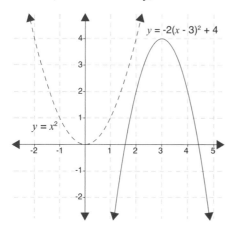

Practice Exercises

Use transformations to sketch the graph.

4.21 $y = -|x+1| - 3$

4.5 Algebra of Functions

Functions can be combined using the basic arithmetic operations of addition, subtraction, multiplication, and division. Furthermore, functions can be composed when one function is used as the input of another function.

Adding, Subtracting, Multiplying, and Dividing Functions

Given two functions f and g, the sum, difference, product, and quotient of those functions are defined as:

- Sum: $(f+g)(x) = f(x) + g(x)$
- Difference: $(f-g)(x) = f(x) - g(x)$
- Product: $(f \cdot g)(x) = f(x) \cdot g(x)$

- Quotient: $\left(\dfrac{f}{g}\right)(x) = \dfrac{f(x)}{g(x)},\ g(x) \neq 0$

The domain of the sum, difference, and product of f and g is the set of values that belong to the domains of both functions f and g.

The domain of the quotient f and g is the set of values that belong to the domains of both functions f and g except for any value of x that makes $g(x) = 0$. Recall that division by 0 is undefined.

Example: If $f(x) = 2x - 1$ and $g(x) = x + 3$, find $(f + g)(x)$, $(f - g)(x)$ and state the domain of each.

Solution:

$(f + g)(x) = f(x) + g(x)$

$\qquad = 2x - 1 + x + 3$

$\qquad = 3x + 2$

The domain of $f(x)$ is all real numbers, and the domain of $g(x)$ is all real numbers. The domain of $(f + g)(x)$ is all real numbers.

$(f - g)(x) = f(x) - g(x)$

$\qquad = 2x - 1 - (x + 3)$

$\qquad = 2x - 1 - x - 3$

$\qquad = x - 4$

The domain of $f(x)$ is all real numbers, and the domain of $g(x)$ is all real numbers. The domain of $(f - g)(x)$ is all real numbers.

Example: If $f(x) = 4x$ and $g(x) = x - 2$, find $(f \cdot g)(x)$, $\left(\dfrac{f}{g}\right)(x)$ and state the domain of each.

Solution:

$(f \cdot g)(x) = f(x) \cdot g(x)$

$\qquad = 4x(x - 2)$

$\qquad = 4x^2 - 8x$

The domain of $f(x)$ is all real numbers, and the domain of $g(x)$ is all real numbers. The domain of $(f \cdot g)(x)$ is all real numbers.

$\left(\dfrac{f}{g}\right)(x) = \dfrac{f(x)}{g(x)},\ g(x) \neq 0$

$\qquad = \dfrac{4x}{x - 2}$

$g(x) = x - 2 = 0$ when $x = 2$.

The domain of $f(x)$ is all real numbers, and the domain of $g(x)$ is all real numbers. The domain of $\left(\dfrac{f}{g}\right)(x)$ is all real numbers except $x = 2$.

Practice Exercises

Find $(f + g)(x)$, $(f - g)(x)$, $(f \cdot g)(x)$, and $\left(\dfrac{f}{g}\right)(x)$ and state any restrictions on the domain for each pair of functions.

4.22 $f(x) = 2x + 3$, $g(x) = x$

4.23 $f(x) = x$, $g(x) = 2x + 3$

> From the Practice Exercises, you can observe that addition and multiplication of functions is commutative, while subtraction and division of functions is not commutative.

Compositions of Functions

Another way that functions can be combined is by composition. When the input of one function is another function, together the functions form a *composite function*.

A composite function $(f \circ g)(x) = f(g(x))$ is read "f of g of x" or "the composition of f with g."

To find $(f \circ g)(x)$, substitute the function g into the function f by substituting the function g for each x that appears in the function f.

Example: If $f(x) = 3x - 2$, and $g(x) = 6x + 4$, find $(f \circ g)(x)$ and $(f \circ g)(4)$

   Solution: Substitute $6x + 4$ for x in the function f.

$$(f \circ g)(x) = f(g(x))$$
$$= f(6x + 4)$$
$$= 3(6x + 4) - 2$$
$$= 18x + 12 - 2$$
$$= 18x + 10$$

You can use the composite function to find the value of $(f \circ g)(4)$.

$(f \circ g)(4) = 18(4) + 10 = 82$

You can also evaluate the functions in sequence to find the value of $(f \circ g)(4)$.

$(f \circ g)(4) = f(g(4))$

$\quad g(4) = 6(4) + 4 = 28$

$\quad f(g(4)) = f(28) = 3(28) - 2 = 82$

When you replace one function by another to form a composite function, make sure you do it in the correct order. Composition of functions is not commutative.

Example: If $f(x) = 3x - 2$, and $g(x) = 6x + 4$, find $(g \circ f)(x)$ and $(g \circ f)(-2)$.

Solution: Substitute $3x - 2$ for x in the function g.

$(g \circ f)(x) = g(f(x))$

$\qquad = g(3x - 2)$

$\qquad = 6(3x - 2) + 4$

$\qquad = 18x - 12 + 4$

$\qquad = 18x - 8$

$(g \circ f)(-2) = 18(-2) - 8 = -44$

The domain of a composite function $(f \circ g)(x)$ is the set of values in the domain of g that produce outputs, or $g(x)$ values, that are in the domain of f. You can see in the first example that when you calculate $(f \circ g)(4)$ using the sequence of functions, the input for g was evaluated and became the output for g. That output became the input for f.

In the previous examples, the domain and range of both functions are all real numbers, so the domain of the composite function is also all real numbers.

Example: What is the domain of $(f \circ g)(x)$ if $f(x) = \dfrac{1}{x+1}$ and

$g(x) = \dfrac{1}{x-2}$?

Solution: First find any restrictions of the domain of g. These values will also be restricted from the domain of the composite function.

Next find restrictions on the domain of f. Find the input values for g that will produce these values, and they must also be restricted from the domain of the composite function.

Restrictions on the domain of g: $x \neq 2$.

Restrictions on the domain of f: $x \neq -1$. So $g(x) \neq -1$.

Solve $g(x) = -1$ to determine the restricted value.

$$\frac{1}{x-2} = -1$$
$$1 = -1(x-2)$$
$$1 = -x + 2$$
$$x = 1$$

The domain of the composite function is all real numbers except 1 and 2.

Practice Exercises

Use the functions $f(x) = \dfrac{1}{x}$ and $g(x) = 2x - 3$.

4.24   Find $(f \circ g)(x)$, $(f \circ g)(2)$ and state the domain of the composite function.

4.25   Find $(g \circ f)(x)$, $(g \circ f)(-2)$ and state the domain of the composite function.

4.6 Inverse Functions

If the elements in the ordered pairs of a relation are reversed, the new set of ordered pairs is the *inverse relation* of the original relation. If R is a relation, then the notation R^{-1} denotes the inverse of R.

Example: Find the inverse of $R = \{(1, 4), (2, 5), (-1, 3), (0, 5)\}$. Graph R, R^{-1} and $y = x$ on the same set of axes.

Solution: To find R^{-1}, reverse the coordinates for each point in R.

$R^{-1} = \{(4, 1), (5, 2), (3, -1), (5, 0)\}$

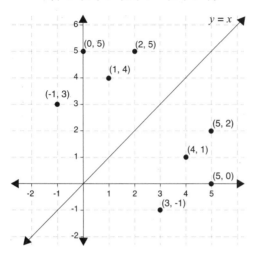

Notice that R and R^{-1} are symmetric about the line $y = x$. In other words, if you folded the graph on the line $y = x$, the points from R and R^{-1} would lie on top of one another.

Notice that R is a function, while R^{-1} is not. Recall that if it is possible to pass a vertical line through two or more points on the graph, then the relation is *not* a function. The inverse of a relation or a function will always be a relation, and possibly, but not always, a function.

To find the inverse of a function, $f^{-1}(x)$, replace $f(x)$ with y, reverse the x and y, and solve for y.

Example: Find the inverse of $f(x) = 2x + 3$. Graph $f(x)$, $f^{-1}(x)$ and $y = x$ on the same set of axes.

Solution: Replace $f(x)$ with y, reverse the x and y and solve for y.

$f(x) = 2x + 3$

$\qquad y = 2x + 3 \qquad$ Replace $f(x)$ with y.

$\qquad x = 2y + 3 \qquad$ Switch x and y.

$x - 3 = 2y \qquad\qquad$ Solve for y.

$$\frac{x-3}{2} = y$$

$$y = \frac{x-3}{2}$$

$$f^{-1}(x) = \frac{x-3}{2}$$

Relations or functions and their inverse will always be symmetric about the line $y = x$. The function $y = x$ or $f(x) = x$ is called the *identity function*.

You can prove that two functions, f and g are *inverse functions* when the composition of f and g and the composition of g and f are both the identity function.

Functions f and g are inverses if and only if $(f \circ g)(x) = (g \circ f)(x) = x$ and $f(g(x)) = g(f(x)) = x$.

Example: Show that $f(x) = 2x + 3$ and $g(x) = \dfrac{x-3}{2}$ are inverses.

Solution: If the functions are inverses, then $(f \circ g)(x) = (g \circ f)(x) = x$.

$(f \circ g)(x) = f(g(x))$ $(g \circ f)(x) = g(f(x))$

$$= 2\left(\frac{x-3}{2}\right) + 3 \qquad\qquad = \frac{(2x+3)-3}{2}$$

$$= x - 3 + 3 \qquad\qquad\qquad = \frac{2x}{2}$$

$$= x \qquad\qquad\qquad\qquad\quad = x$$

Since $(f \circ g)(x) = (g \circ f)(x) = x$, $f(x)$ and $g(x)$ are inverses.

For all linear functions except horizontal or vertical lines, the inverse will always be a function. For many nonlinear functions, their inverses are not functions.

Example: If $f(x) = x^2$, find the inverse and graph the function, with the inverse and $y = x$ on the same graph. Is the inverse a function?

Solution: Replace $f(x)$ with y, reverse the x and y, and solve for y.

$f(x) = x^2$

$\quad y = x^2$ $\qquad$ Replace $f(x)$ with y.

$\quad x = y^2$ $\qquad$ Switch x and y.

$\quad y^2 = x$

$\quad y = \pm\sqrt{x}$ $\qquad$ Solve for y.

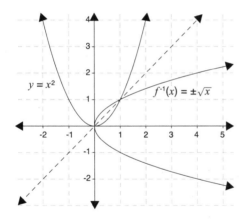

The inverse is not a function because you can pass a vertical line through the inverse and touch two points. The inverse fails the vertical line test for a function.

Practice Exercises

4.26   Determine whether the functions are inverses of each other: $f(x) = 3x - 1$, $g(x) = x + 3$.

4.27   Find the inverse of $f(x) = 6x + 2$.

Chapter 5

Quadratic Functions and Relations

5.1 Basic Definitions

A *quadratic function* has the form $f(x) = ax^2 + bx + c$ where $a \neq 0$. Notice that a quadratic function has a quadratic term, a linear term and a constant term.

$$f(x) = \underbrace{ax^2}_{\text{quadratic}} + \underbrace{bx}_{\text{linear}} + \underbrace{c}_{\text{constant}}$$

The *general form* of a *quadratic equation* is $ax^2 + bx + c = 0$ where $a \neq 0$.

The solution to a quadratic equation can be represented graphically by where the corresponding quadratic function crosses the *x*-axis, where $f(x) = 0$. A quadratic equation can have either two real solutions, one real solution, or no real solutions (Figure 5.1). The U-shaped graph of a quadratic function is called a parabola.

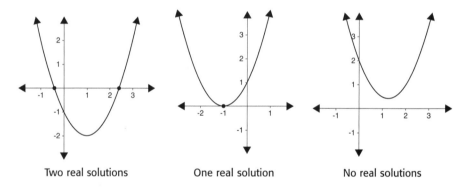

Two real solutions   One real solution   No real solutions

Figure 5.1 Solutions to a quadratic equation

When the graph of a quadratic function crosses the *x*-axis in two places, there are two real solutions to the corresponding quadratic equation. When the graph of a quadratic function touches the *x*-axis in only one place, there is only one real solution to the corresponding quadratic equation. When the graph of a quadratic function does not cross the *x*-axis, there are no real solutions to the corresponding quadratic equation.

The graph of a quadratic function can be helpful in verifying the number and type of solution(s) or to estimate or check a solution to a quadratic equation, but it is not the most efficient or practical method for solving a quadratic equation. We will explore several methods for graphing a quadratic function in Lesson 5.7.

There are several methods for solving a quadratic equation algebraically, including factoring, completing the square, and applying the quadratic formula.

5.2 Solving Quadratic Equations by Factoring

Solving a quadratic equation by factoring relies on the *Zero Product Property,* which states that if a product equals zero, then at least one of the factors is equal to zero.

Zero Product Property

If $ab = 0$, then $a = 0$ or $b = 0$

Common Factors

In order to use the Zero Product Property, the quadratic equation must be rewritten as the product of two factors. If $c = 0$ in the quadratic equation $ax^2 + bx + c = 0$, then factor out the common factor.

Example: Solve $5x^2 - 20x = 0$

> **Solution:** Both terms have a common factor of $5x$. Factor out the common factor and solve using the Zero Product Property.

$5x^2 - 20x = 0$

$5x(x - 4) = 0$		Factor $5x$ from both terms.
$5x = 0$ or $x - 4 = 0$		Use Zero Product Property.
$x = 0$ or $x = 4$		Solve for x.

Check:

$$5(0)^2 - 20(0) = 0 \qquad\qquad 5(4)^2 - 20(4) = 0$$

$$0 - 0 = 0 \checkmark \qquad\qquad 80 - 80 = 0 \checkmark$$

0 and 4 are solutions to the quadratic equation.

Factor Patterns and the Square Root Property

In the previous example, the terms had a common factor that could be factored out in order to use the Zero Product Property. In many cases, there will be no common factor, and a factor pattern can help you factor the quadratic equation.

Factor Patterns

- Difference of two perfect squares

       General Form $a^2 - b^2 = (a+b)(a-b)$

       Example $4x^2 - 9 = (2x+3)(2x-3)$

- Perfect square trinomial

       General Form $a^2 + 2ab + b^2 = (a+b)^2$ $a^2 - 2ab + b^2 = (a-b)^2$

       Example $16x^2 + 24x + 9 = (4x+3)^2$ $4x^2 - 20x + 25 = (2x-5)^2$

Example: Solve $x^2 - 36 = 0$

     Solution: $x^2 - 36$ is the difference of two perfect squares. Factor and use the Zero Product Property.

$$x^2 - 36 = 0$$

$$(x+6)(x-6) = 0 \qquad \text{Difference of two perfect squares factor pattern.}$$

$$x + 6 = 0 \quad \text{or} \quad x - 6 = 0 \qquad \text{Use Zero Product Property.}$$

$$x = -6 \quad \text{or} \qquad x = 6 \qquad \text{Solve for } x.$$

Check:

$$(-6)^2 - 36 = 0 \qquad\qquad (6)^2 - 36 = 0$$

$$36 - 36 = 0 \checkmark \qquad\qquad 36 - 36 = 0 \checkmark$$

     −6 and 6 are solutions to the quadratic equation.

The previous example can also be solved using the *Square Root Property*.

Square Root Property

If $x^2 = c$, then $x = \sqrt{c}$ or $x = -\sqrt{c}$

Or,

If $x^2 = c$, then $x = \pm\sqrt{c}$

$x^2 - 36 = 0$ becomes $x^2 = 36$, and using the Square Root Property,
$x = \pm\sqrt{36} = \pm 6$.

The Square Root Property is even simpler if the perfect square is equal to zero.

Example: Solve $4x^2 + 20x + 25 = 0$

Solution: Notice that $4x^2$ and 25 are perfect squares. In the perfect square trinomial pattern, $a = 2x$, $b = 5$ and $2ab = 20x$.

$4x^2 + 20x + 25 = 0$

$\qquad (2x + 5)^2 = 0$ Perfect square trinomial factor pattern.

$\qquad 2x + 5 = 0$ Use the Square Root Property.

$\qquad x = -\dfrac{5}{2}$ Solve.

Check:

$$4\left(-\frac{5}{2}\right)^2 + 20\left(-\frac{5}{2}\right) + 25 = 0$$

$$25 - 50 + 25 = 0 \checkmark$$

$-\dfrac{5}{2}$ is the solution to the quadratic equation.

If a trinomial does not fit a factor pattern, you can use a different technique to try to factor. In the trinomial $ax^2 + bx + c$, find ac. Identify whether there is a set of factors of ac whose sum is b.

Consider $6x^2 - 5x - 4$. $ac = 6 \cdot -4 = -24$.

Factors of –24	Sum of Factors	Factors of –24	Sum of Factors
1, –24	–23	–1, 24	23
2, –12	–10	–2, 12	10
3, –8	–5	–3, 8	5
4, –6	–2	–4, 6	2

The factors 3 and –8 give the sum $-5 = b$. Rewrite the trinomial, splitting the middle term into the sum of the two factors, and then factor by grouping.

$6x^2 - 5x - 4$

$6x^2 + 3x + -8x - 4$ Rewrite middle term as the sum of the factors.

$3x(2x + 1) - 4(2x + 1)$ Factor by grouping.

$(3x - 4)(2x + 1)$ Factor by grouping.

Example: Solve $4x^2 - 13x + 3 = 0$

Solution: The trinomial is not a perfect square trinomial. Identify ac and find the factors of ac whose sum is –13.

$ac = 12$

–12 and –1 are factors of ac that add to –13.

$4x^2 - 13x + 3 = 0$

$4x^2 - 12x - 1x + 3 = 0$ Split middle term into factors of ac.

$4x(x - 3) - 1(x - 3) = 0$ Factor by grouping.

$(4x - 1)(x - 3) = 0$ Factor by grouping.

$4x - 1 = 0$ or $x - 3 = 0$ Use Zero Product Property.

$x = \dfrac{1}{4}$ or $x = 3$ Solve for x.

Check:

$$4\left(\frac{1}{4}\right)^2 - 13\left(\frac{1}{4}\right) + 3 = 0 \qquad\qquad 4(3)^2 - 13(3) + 3 = 0$$

$$\frac{1}{4} - \frac{13}{4} + \frac{12}{4} = 0 \checkmark \qquad\qquad 36 - 39 + 3 = 0 \checkmark$$

$\frac{1}{4}$ and 3 are solutions to the quadratic equation.

Sometimes a quadratic equation that looks difficult can be simplified by factoring out a common factor.

Example: Solve $20x^2 + 60x + 45 = 0$

> **Solution:** A common factor is 5. Factor 5 out of the trinomial and then factor what remains.
>
> $20x^2 + 60x + 45 = 0$
>
> $5(4x^2 + 12x + 9) = 0$ Factor out the common factor.
>
> $5(2x + 3)^2 = 0$ Perfect square trinomial factor pattern.
>
> $(2x + 3)^2 = 0$ Divide both sides of the equation by 5.
>
> $2x + 3 = 0$ Use square root property.
>
> $x = -\dfrac{3}{2}$ Solve for x.
>
> Check:
>
> $$20\left(-\frac{3}{2}\right)^2 + 60\left(-\frac{3}{2}\right) + 45 = 0$$
>
> $$45 - 90 + 45 = 0 \checkmark$$
>
> $-\dfrac{3}{2}$ is the solution to the quadratic equation.

Strategies for Factoring Quadratic Equations in General Form
- Factor out any common factors.
- Look for the difference of two perfect squares or perfect square trinomial factor patterns.
- Calculate ac and identify whether there is a set of factors of ac whose sum is b. If so, rewrite bx using the sum of the factors of ac.

Practice Exercises

Solve each quadratic equation by factoring.

5.1 $15x^2 - 6x = 0$

5.2 $16x^2 - 36 = 0$

5.3 $4x^2 + 4x + 1 = 0$

5.4 $2x^2 - 5x - 12 = 0$

5.5 $15x^2 + 39x - 18 = 0$

5.3 Completing the Square

In Lesson 5.2 you solved quadratic equations by factoring. However, many quadratic equations are not factorable, so other techniques are needed.

When you *complete the square*, you will make a perfect square trinomial on one side of the equation and then solve the equation by using the Square Root Property.

Steps for Completing the Square for $ax^2 + bx + c = 0$, $a \neq 0$

- If necessary, transform the equation so that the coefficient of x^2 is 1.

- Isolate c on the right side of the equation by adding $-c$ to both sides of the equation.

- Multiply b by $\dfrac{1}{2}$ and then square the result.

- Add $\left(\dfrac{b}{2}\right)^2$ to both sides of the equation.

- Factor the left side of the equation into a perfect square trinomial.

- Use the Square Root Property.

Example: Solve $x^2 + 14x + 1 = 0$ by completing the square.

Solution:

$x^2 + 14x + 1 = 0$	The coefficient of x^2 is already 1.
$x^2 + 14x = -1$	Add -1 to both sides.
$x^2 + 14x + 49 = -1 + 49$	$\left(\dfrac{14}{2}\right)^2 = 49$. Add 49 to both sides.
$(x + 7)^2 = 48$	Factor the perfect square trinomial.
$x + 7 = \pm\sqrt{48}$	Use the Square Root Property.
$x = -7 \pm \sqrt{48}$	Solve for x.

Recall that if the number under the square root symbol has a factor that is a perfect square, the square root can be simplified.

$$\sqrt{48} = \sqrt{16 \cdot 3} = \sqrt{16} \cdot \sqrt{3} = 4\sqrt{3}$$

$$x = -7 \pm 4\sqrt{3}$$

You can check the irrational solution to a quadratic equation by estimation or by using the exact answer. To check an irrational solution by estimation, calculate the value of the solution and round the number.

From the previous example, the solutions to $x^2 + 14x + 1 = 0$ are

$-7 + 4\sqrt{3} \approx -0.0718$ and $-7 - 4\sqrt{3} \approx -13.9282$.

Check:

$(-0.0718)^2 + 14(-0.0718) + 1 \approx 0 \qquad (-13.9282)^2 + 14(-13.9282) + 1 \approx 0$

$0.0052 - 1.0052 + 1 \approx 0 \checkmark \qquad\qquad 193.9948 - 194.9948 + 1 \approx 0 \checkmark$

To check the solution with the exact irrational solution, you must treat a number such as $-7 \pm 4\sqrt{3}$ like a binomial. Squaring the irrational solution creates a perfect square trinomial.

$$\left(-7 + 4\sqrt{3}\right)^2 = (-7)^2 + 2\left(-28\sqrt{3}\right) + \left(4\sqrt{3}\right)^2$$

$$= 49 - 56\sqrt{3} + 48$$

$$= 97 - 56\sqrt{3}$$

From the previous example, the solutions to $x^2 + 14x + 1 = 0$ are $-7 \pm 4\sqrt{3}$.

Check $-7 + 4\sqrt{3}$:

$$\left(-7 + 4\sqrt{3}\right)^2 + 14\left(-7 + 4\sqrt{3}\right) + 1 = 0$$

$$97 - 56\sqrt{3} - 98 + 56\sqrt{3} + 1 = 0$$

$$97 - 98 + 1 - 56\sqrt{3} + 56\sqrt{3} = 0 \checkmark$$

Checking the other solution, $-7 - 4\sqrt{3}$, will be left to you.

Example: Solve $2x^2 - 5x - 4 = 0$ by completing the square.

Solution:

$2x^2 - 5x - 4 = 0$

$x^2 - \dfrac{5}{2}x - 2 = 0$ Divide both sides by 2.

$x^2 - \dfrac{5}{2}x = 2$ Add 2 to both sides.

$x^2 - \dfrac{5}{2}x + \dfrac{25}{16} = 2 + \dfrac{25}{16}$ $\left(-\dfrac{5}{2} \cdot \dfrac{1}{2}\right)^2 = \dfrac{25}{16}$. Add $\dfrac{25}{16}$ to both sides.

$\left(x - \dfrac{5}{4}\right)^2 = \dfrac{57}{16}$ Factor the perfect square trinomial.

$x - \dfrac{5}{4} = \pm\sqrt{\dfrac{57}{16}}$ Use the Square Root Property.

$x = \dfrac{5}{4} \pm \sqrt{\dfrac{57}{16}}$ Solve for x.

$x = \dfrac{5}{4} \pm \dfrac{\sqrt{57}}{4}$ Simplify.

$x = \dfrac{5 \pm \sqrt{57}}{4}$ Simplify.

The check is left to you.

Practice Exercises

Solve each quadratic equation by completing the square.

5.6　$x^2 + 10x + 20 = 0$

5.7　$3x^2 - 9x - 27 = 0$

5.4 Complex Numbers

So far, you have solved quadratic equations that have one or two real solutions. Recall from Lesson 5.1 that some quadratic equations have no real solutions. That is, there are no real numbers that are solutions.

Consider the quadratic equation $x^2 + 1 = 0$.

$$x^2 + 1 = 0$$
$$x^2 = -1 \qquad \text{Add } -1 \text{ to both sides.}$$
$$x = \pm\sqrt{-1} \qquad \text{Use the Square Root Property.}$$

Previously, you learned that the square root of a negative number is undefined in the real number system, so the quadratic equation $x^2 + 1 = 0$ has no real solution. However, to address this issue, the *imaginary numbers* were invented. The imaginary number $i = \sqrt{-1}$. So, the quadratic equation $x^2 + 1 = 0$ has two imaginary solutions, $\pm i$.

Working with Imaginary Numbers

Imaginary Numbers

$$i = \sqrt{-1}$$

$$i^2 = -1$$

Square roots of negative numbers can be rewritten with i. Split the square of the negative number into the product of the square root of -1 and the square of a positive number. Substitute i for the square root of -1.

Example: Simplify each square root: $\sqrt{-5}$, $\sqrt{-16}$, $\sqrt{-12}$

　　Solution: Factor $\sqrt{-1}$ from each square root and substitute i for $\sqrt{-1}$. Simplify the remaining square root if possible.

$$\sqrt{-5} = \sqrt{-1} \cdot \sqrt{5} = i\sqrt{5}$$

$$\sqrt{-16} = \sqrt{-1} \cdot \sqrt{16} = 4i$$

$$\sqrt{-12} = \sqrt{-1} \cdot \sqrt{12} = \sqrt{-1} \cdot \sqrt{4} \cdot \sqrt{3} = 2i\sqrt{3}$$

The commutative and associative properties are true for imaginary numbers. The successive powers of i illustrate a pattern.

$i^1 = i$ $i^5 = i^4 \cdot i = 1 \cdot i = i$

$i^2 = -1$ $i^6 = i^4 \cdot i^2 = 1 \cdot -1 = -1$

$i^3 = i^2 \cdot i = -1i = -i$ $i^7 = i^4 \cdot i^3 = 1 \cdot -i = -i$

$i^4 = i^2 \cdot i^2 = -1 \cdot -1 = 1$ $i^8 = i^4 \cdot i^4 = 1 \cdot 1 = 1$

Express the square root of any negative number in terms of i before simplifying an expression.

Example: Simplify i^{14}, $3i \cdot 6i$, $\sqrt{-2} \cdot \sqrt{-6}$

Solution:

$i^{14} = i^7 \cdot i^7 = -i \cdot -i = i^2 = -1$

$3i \cdot 6i = 18i^2 = 18 \cdot -1 = -18$

$\sqrt{-2} \cdot \sqrt{-6} = i\sqrt{2} \cdot i\sqrt{6} = i^2\sqrt{12} = -\sqrt{4} \cdot \sqrt{3} = -2\sqrt{3}$

Practice Exercises

Simplify.

5.8 $\sqrt{-243}$

5.9 i^{25}

5.10 $2i \cdot (-7i)$

5.11 $\sqrt{-12} \cdot \sqrt{-3}$

Solving Quadratic Equations with Complex Solutions

Complex numbers are the set of real and imaginary numbers. Complex numbers have the form $a + bi$, where a and b are real numbers. Real numbers can be expressed in this system by $a + 0i$. Complex numbers can be used to solve quadratic equations that do not have real solutions.

Example: Solve $x^2 + 6x + 21 = 0$

Solution: The trinomial cannot be factored since it does not fit a factor pattern and there are no two factors of 21 whose sum is 6. Complete the square to solve.

$x^2 + 6x + 21 = 0$

$x^2 + 6x = -21$	Add –21 to both sides of the equation.
$x^2 + 6x + 9 = -21 + 9$	$\left(\dfrac{6}{2}\right)^2 = 9$. Add 9 to both sides of the equation.
$(x + 3)^2 = -12$	Factor the perfect square trinomial.
$x + 3 = \pm\sqrt{-12}$	Use the Square Root Property.
$x = -3 \pm \sqrt{-12}$	Solve for x.
$x = -3 \pm \sqrt{-1} \cdot \sqrt{4} \cdot \sqrt{3}$	Simplify.
$x = -3 \pm 2i\sqrt{3}$	Simplify to a complex number.

Check $-3 + 2i\sqrt{3}$:

$$\left(-3 + 2i\sqrt{3}\right)^2 + 6\left(-3 + 2i\sqrt{3}\right) + 21 = 0$$

$$\left(-3\right)^2 + 2(-3)\left(2i\sqrt{3}\right) + \left(2i\sqrt{3}\right)^2 - 18 + 12i\sqrt{3} + 21 = 0$$

$$9 - 12i\sqrt{3} - 12 - 18 + 12i\sqrt{3} + 21 = 0 \checkmark$$

The check for the other solution is left to you.

Notice that the quadratic equation in the previous example has no real solutions; however, it does have two complex solutions.

Practice Exercises

Solve.

5.12 $x^2 + 20 = 0$

5.13 $x^2 - 8x + 25 = 0$

5.14 $x^2 + 5x + 7 = 0$

5.5 Solving Quadratic Equations by the Quadratic Formula

Instead of completing the square for each quadratic equation you encounter, you can utilize the quadratic formula, which is the general formula for completing the square for a quadratic equation in general form.

Quadratic Formula

For a quadratic equation of the form $ax^2 + bx + c = 0$, where a, b, and c are real numbers, and $a \neq 0$,

$$x = \frac{-b \pm \sqrt{b^2 - 4ac}}{2a}$$

The quadratic formula can be used to find real and complex solutions to quadratic equations.

Example: Solve $x^2 - 4x + 2 = 0$ using the quadratic formula.

Solution: Identify a, b, and c, substitute them into the quadratic formula, and simplify.

$a = 1$, $b = -4$, $c = 2$

$$x = \frac{-(-4) \pm \sqrt{(-4)^2 - 4(1)(2)}}{2(1)}$$

$$x = \frac{4 \pm \sqrt{16 - 8}}{2}$$

$$x = \frac{4 \pm \sqrt{8}}{2} = \frac{4 \pm \sqrt{4} \cdot \sqrt{2}}{2} = \frac{4 \pm 2\sqrt{2}}{2} = 2 \pm \sqrt{2}$$

$2 \pm \sqrt{2}$ are solutions to the quadratic equation.

Check $2 + \sqrt{2}$:

$$\left(2 + \sqrt{2}\right)^2 - 4\left(2 + \sqrt{2}\right) + 2 = 0$$

$$2^2 + 2\left(2\sqrt{2}\right) + \left(\sqrt{2}\right)^2 - 8 - 4\sqrt{2} + 2 = 0$$

$$4 + 4\sqrt{2} + 2 - 8 - 4\sqrt{2} + 2 = 0 \checkmark$$

The check for the other solution is left to you.

Solutions obtained by the quadratic formula can be checked using exact or estimated values.

Practice Exercises

Solve using the quadratic formula.

5.15 $x^2 + 5x - 3 = 0$

5.16 $2x^2 - 7x + 14 = 0$

Comparing Solution Methods

In this chapter you have learned three methods for solving quadratic equations: factoring, completing the square, and the quadratic formula. Consider the methods side by side.

Factoring	Completing the Square	Quadratic Formula
$x^2 + 3x + 2 = 0$	$x^2 + 3x + 2 = 0$	$x^2 + 3x + 2 = 0$
$(x + 2)(x + 1) = 0$	$x^2 + 3x = -2$	$x = \dfrac{-3 \pm \sqrt{3^2 - 4(1)(2)}}{2(1)}$
$x + 2 = 0$ or $x + 1 = 0$	$x^2 + 3x + \dfrac{9}{4} = -2 + \dfrac{9}{4}$	$x = \dfrac{-3 \pm \sqrt{9 - 8}}{2}$
$x = -2$ or $x = -1$	$\left(x + \dfrac{3}{2}\right)^2 = \dfrac{1}{4}$	$x = \dfrac{-3 \pm \sqrt{1}}{2}$
	$x + \dfrac{3}{2} = \pm\sqrt{\dfrac{1}{4}}$	$x = \dfrac{-3 \pm 1}{2}$
	$x = -\dfrac{3}{2} \pm \dfrac{1}{2}$	$x = -2$ or -1
	$x = -2$ or -1	

Factoring is quick and simple, but not all quadratic expressions can be factored. Completing the square is fairly straightforward if b is even. If b is odd, completing the square will introduce fractions. The quadratic formula is favored by some because it requires only simplifying a numeric expression, and it will generate a solution to any quadratic equation.

Practice Exercises

Solve the quadratic equation by the method of your choice.

5.17 $x^2 - 6x + 8 = 0$

5.18 $x^2 + 10x - 3 = 0$

5.19 $x^2 - 5x + 9 = 0$

The Discriminant

In the quadratic formula $x = \dfrac{-b \pm \sqrt{b^2 - 4ac}}{2a}$, the expression under the

square root symbol, $b^2 - 4ac$, is called the *discriminant*. The discriminant can be used to identify the kind of solution the quadratic equation will have.

If $b^2 - 4ac > 0$, the equation has two real solutions.

If $b^2 - 4ac = 0$, the equation has one real solution.

If $b^2 - 4ac < 0$, the equation has two complex solutions.

Example: Determine the number and kind of solutions for each quadratic equation.

$$x^2 - 7x + 10 = 0 \qquad\qquad -2x^2 + x - 5 = 0$$

Solution: Substitute a, b, and c into the discriminant $b^2 - 4ac$.

$x^2 - 7x + 10 = 0 : a = 1,\ b = -7,\ c = 10$

$b^2 - 4ac = \left(-7\right)^2 - 4(1)(10) = 9$

$9 > 0$, so there are two real solutions.

$-2x^2 + x - 5 = 0 : a = -2,\ b = 1,\ c = -5$

$b^2 - 4ac = 1^2 - 4(-2)(-5) = -39$

$-39 < 0$, so there are two complex solutions.

Practice Exercises

Identify the number and kind of solutions for each quadratic equation.

5.20 $x^2 + 4x - 10 = 0$

5.21 $3x^2 - 5x + 6 = 0$

5.6 Solving Quadratic Inequalities

Quadratic inequalities are of the form $ax^2 + bx + c < 0$ or $ax^2 + bx + c > 0$ where $a \neq 0$.

To find the real solutions to a quadratic inequality:

- Solve the related quadratic equation.

- The solutions to the equation become boundary points that divide the number line into regions.

- Select points from each region created by the boundary points and determine whether they make the original inequality true.

- Regions that contain a point that satisfies the inequality are part of the solution. Boundary points are part of the solution if the inequality is $\leq$ or $\geq$. On a graph, use open or closed dots to indicate whether the boundary points are included in the solution.

Example: Solve $(x + 3)(x - 2) < 0$

> **Solution:** Rewrite the inequality as an equation and solve to identify the boundary points.
>
> $(x + 3)(x - 2) = 0$ Related equation.
>
> $x + 3 = 0$ or $x - 2 = 0$ Use the Zero Product Property.
>
> $x = -3$ or $x = 2$ Solve for x.
>
> The boundary points create 3 regions:
>
>
> Choose points from each region and substitute into the inequality:
>
> | $x = -4$ | $x = 0$ | $x = 3$ |
> |---|---|---|
> | $(-4 + 3)(-4 - 2) < 0$ | $(0 + 3)(0 - 2) < 0$ | $(3 + 3)(3 - 2) < 0$ |
> | $(-1)(-6) < 0$ | $(3)(-2) < 0$ | $(6)(1) < 0$ |
> | $6 < 0$ | $-6 < 0 \checkmark$ | $6 < 0$ |

The region that satisfies in the inequality is $-3 < x < 2$. The boundary points are not included because the inequality is $<$ and not $\leq$.

The solution to the inequality is $-3 < x < 2$.

Solutions to inequalities can be given in several different forms. Solutions to inequalities in Chapter 1 were shown in graph form. Solutions may also be shown in set form or interval form.

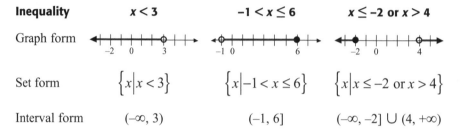

Inequality	$x < 3$	$-1 < x \leq 6$	$x \leq -2$ or $x > 4$
Graph form			
Set form	$\{x \mid x < 3\}$	$\{x \mid -1 < x \leq 6\}$	$\{x \mid x \leq -2 \text{ or } x > 4\}$
Interval form	$(-\infty, 3)$	$(-1, 6]$	$(-\infty, -2] \cup (4, +\infty)$

Set form simply puts the inequality in a set and denotes the variable to the left of the vertical line.

Interval form indicates the left and right boundary points of the region separated by a comma. A bracket indicates that the boundary point is included in the region, while a parenthesis indicates that the boundary point is not included in the region. The bracket and parenthesis are analogous to the closed and open dots on a graph. To indicate that the region is unbounded in the positive or negative direction, the positive or negative infinity symbol is used: $+\infty$ or $-\infty$. An infinity symbol will always be enclosed with a parenthesis, not a bracket. If two regions are solutions, the two intervals will be joined by $\cup$, the union symbol.

Example: Solve $2x^2 - 13x + 15 \geq 0$

Solution: Rewrite the inequality as an equation and solve by a method of your choice.

$2x^2 - 13x + 15 = 0$ Related equation.

$(2x - 3)(x - 5) = 0$ Factor the trinomial.

$2x - 3 = 0$ or $x - 5 = 0$ Use the Zero Product Property.

$x = \dfrac{3}{2}$ or $x = 5$ Simplify.

The boundary points create 3 regions:

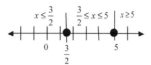

Choose points from each region and substitute into the inequality:

$x = 0$ $x = 2$ $x = 6$

$2(0)^2 - 13(0) + 15 \geq 0$ $2(2)^2 - 13(2) + 15 \geq 0$ $2(6)^2 - 13(6) + 15 \geq 0$

$\qquad\qquad 15 \geq 0 \checkmark$ $-3 \geq 0$ $9 \geq 0 \checkmark$

The solution to the quadratic inequality is $x \leq \dfrac{3}{2}$ or $x \geq 5.$

In set form, the solution is $\left\{ x \middle| x \leq \dfrac{3}{2} \text{ or } x \geq 5 \right\}.$

In interval form, the solution is $\left(-\infty,\ \dfrac{3}{2} \right] \cup [5,\ +\infty).$

Recall that you can solve quadratic equations by factoring, completing the square, and using the quadratic formula. If you encounter irrational solutions, approximate their value to identify the boundaries and solution regions.

Example: Solve $x^2 + 2 < 6x$

Solution: Rewrite the inequality as an equation. Transform the equation so it equals 0 and solve the quadratic equation by a method of your choice.

$x^2 + 2 = 6x$ Related equation.

$x^2 - 6x + 2 = 0$ Subtract $6x$ from both sides.

$x = \dfrac{-(-6) \pm \sqrt{(-6)^2 - 4(1)(2)}}{2(1)}$ Use the quadratic equation.

$x = \dfrac{6 \pm \sqrt{36 - 8}}{2}$ Simplify.

$$x = \frac{6 \pm \sqrt{28}}{2} \qquad \text{Simplify.}$$

$$x = \frac{6 \pm 2\sqrt{7}}{2} = 3 \pm \sqrt{7} \qquad \text{Simplify.}$$

$x \approx 5.6458$ or $x \approx 0.3542$

The boundary points create 3 regions:

Choose points from each region and substitute into the inequality:

$x = 0$	$x = 1$	$x = 6$
$(0)^2 + 2 < 6(0)$	$(1)^2 + 2 < 6(1)$	$(6)^2 + 2 < 6(6)$
$2 < 0$	$3 < 6$ ✓	$38 < 36$

The solution to the quadratic inequality is $3 - \sqrt{7} < x < 3 + \sqrt{7}$.

In set form, the solution is $\left\{ x \middle| 3 - \sqrt{7} < x < 3 + \sqrt{7} \right\}$.

In interval form, the solution is $\left(3 - \sqrt{7},\ 3 + \sqrt{7} \right)$.

Practice Exercises

Solve.

5.22 $2x^2 - 7x - 4 \leq 0$

5.23 $x^2 + 5x > 1$

5.7 Graphing Quadratic Functions

The U-shaped graph of a quadratic function is called a parabola. There are several different techniques you can use to graph a parabola.

Using a Table of Values

Recall that ordered pairs that satisfy the function lie on the graph of the function. To graph a quadratic equation using a table of values, choose values for x, evaluate the function, and graph the coordinates.

Example: Graph $f(x) = 2x^2 - 12x + 10$

> **Solution:** Choose values for x, evaluate the function, graph the coordinates, and connect the coordinates with a smooth curve.

x	$2x^2 - 12x + 10$	$f(x)$	$(x, f(x))$
0	$2(0)^2 - 12(0) + 10$	10	$(0, 10)$
1	$2(1)^2 - 12(1) + 10$	0	$(1, 0)$
2	$2(2)^2 - 12(2) + 10$	-6	$(2, -6)$
3	$2(3)^2 - 12(3) + 10$	-8	$(3, -8)$
4	$2(4)^2 - 12(4) + 10$	-6	$(4, -6)$
5	$2(5)^2 - 12(5) + 10$	0	$(5, 0)$
6	$2(6)^2 - 12(6) + 10$	10	$(6, 10)$

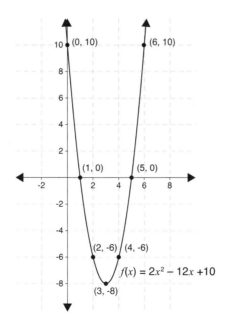

Practice Exercises

Graph each quadratic function using a table of values.

5.24 $f(x) = x^2 - 6x + 10$

5.25 $f(x) = -2x^2 - 4x + 5$

Using General Form

The U-shaped graph of the quadratic function displays a *line of symmetry*. The line of symmetry of a parabola divides the parabola in half and crosses the parabola at only one point, the *vertex*.

In the parabola at the right, the axis of symmetry is $x = -1$, and the vertex is $(-1, 2)$. Points $(0, 3)$ and $(-2, 3)$ demonstrate the symmetry of a parabola. Both points are one unit away from the axis of symmetry. Every point on the parabola except the vertex has a corresponding point that is the same distance from the line of symmetry. Notice that the y-intercept of the graph is 3.

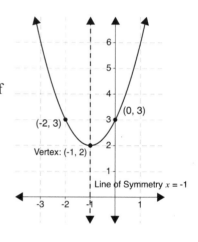

The function represented by the graph at the right is $f(x) = x^2 + 2x + 3$.

Facts about the line of symmetry, vertex, and the y-intercept can all be gleaned from a quadratic function in general form.

For a quadratic function $f(x) = ax^2 + bx + c$, $a \neq 0$, the following is true:

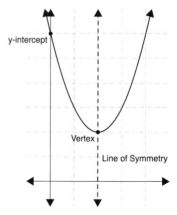

- The line of symmetry is given by the line $x = \dfrac{-b}{2a}$.

- The x-coordinate of the vertex is $\dfrac{-b}{2a}$.
- The y-intercept $= c$.

To sketch the graph of a quadratic function in general form,

- Identify and sketch the line of symmetry.
- Identify the x-coordinate of the vertex and evaluate the function to find the coordinates of the vertex. Plot the vertex.
- Plot the y-intercept and its reflection on the other side of the line of symmetry.
- Draw a smooth curve through the points.

Example: Graph $f(x) = 2x^2 - 12x + 10$

> **Solution:** Identify the line of symmetry, vertex, y-intercept, and its reflection. Draw a smooth curve through the points.
>
> Line of symmetry: $x = \dfrac{-b}{2a} = \dfrac{-(-12)}{2(2)} = 3$
>
> x-coordinate of vertex: $\dfrac{-b}{2a} = \dfrac{-(-12)}{2(2)} = 3$
>
> y-coordinate of vertex: $f(3) = 2(3)^2 - 12(3) + 10 = -8$
>
> vertex: $(3, -8)$
>
> y-intercept: $c = 10$

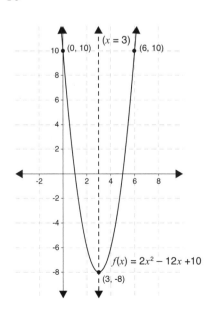

Practice Exercises

Sketch a graph of each quadratic function.

5.26 $f(x) = x^2 + 4x + 1$

5.27 $f(x) = 2x^2 - 8x + 5$

Using Vertex Form

In Lesson 4.4, you learned how to graph quadratic functions of the form $y = a(x - h)^2 + k$, $a \neq 0$ by considering transformations to the parent function $f(x) = x^2$.

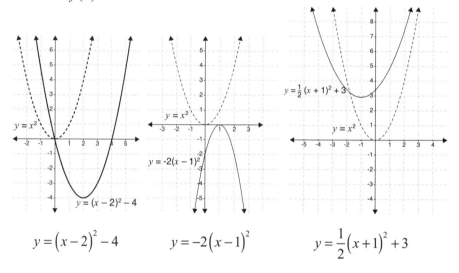

$$y = (x-2)^2 - 4 \qquad y = -2(x-1)^2 \qquad y = \frac{1}{2}(x+1)^2 + 3$$

$y = a(x - h)^2 + k$ is called *vertex form* because translations by h and k identify the vertex (h, k) of the parabola. The line of symmetry of the parabola is $x = h$. Recall from Lesson 4.4 that the value of a determines the direction and shape of the parabola. If $a < 0$, the parabola is reflected over the x-axis. If $0 < |a| < 1$, the parabola is compressed vertically, and if $|a| > 1$ the parabola is stretched vertically.

You can sketch the parabola using the information from the vertex form. If you need a more precise graph, you can choose several values for x and evaluate the function to find additional coordinates on the parabola.

Example: Graph the quadratic function $y = -3(x+4)^2 - 2$

Solution: Identify the vertex and line of symmetry. Evaluate a to determine the shape and direction of the parabola. Find additional coordinates by choosing values for x and evaluating the function to find y.

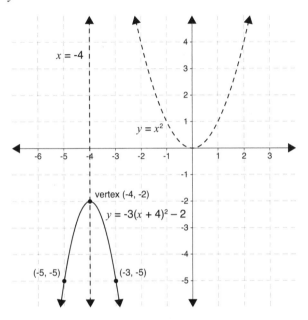

$h = -4$, $k = -2$, so the vertex is $(-4, -2)$

The line of symmetry is $x = -4$.

$a < 0$, so the parabola is reflected over the x-axis and opens down.

$|a| > 1$, so the parabola is stretched vertically.

When $x = -3$, $y = -3(-3+4)^2 - 2 = -5$, so $(-3, -5)$ is on the parabola.

When $x = -5$, $y = -3(-5+4)^2 - 2 = -5$, so $(-5, -5)$ is on the parabola.

You can transform quadratic functions in general form, $f(x) = ax^2 + bx + c$, to vertex form by completing the square.

Example: Write $y = x^2 + 2x + 7$ in vertex form and graph.

Solution: Write the quadratic function in vertex form by completing the square. Identify the vertex and the line of symmetry. Sketch the parabola.

$$y = x^2 + 2x + 7$$

$$y = x^2 + 2x + (1 - 1) + 7 \qquad \left(\frac{2}{2}\right)^2 = 1. \quad \text{Add and subtract 1 from the right side.}$$

$$y = (x^2 + 2x + 1) + 6 \qquad \text{Regroup and simplify.}$$

$$y = (x + 1)^2 + 6 \qquad \text{Factor the perfect square trinomial.}$$

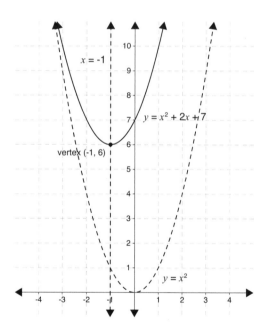

$h = -1$ and $k = 6$, so the vertex of the parabola is $(-1, 6)$.

The line of symmetry is $x = -1$.

$a = 1$, so the shape and direction of the parabola is the same as the parent function $f(x) = x^2$.

In general form, $y = ax^2 + bx + c$, if $a \neq 1$, factor a from the quadratic and linear terms before completing the square.

Example: Write $y = 2x^2 - 12x + 10$ in vertex form and graph.

Solution: Factor 2 from the quadratic and linear terms. Write the quadratic function in vertex form by completing the square. Identify the vertex and the line of symmetry. Sketch the parabola.

$$y = 2x^2 - 12x + 10$$

$$y = 2(x^2 - 6x) + 10 \qquad\qquad \text{Factor 2 from quadratic}$$
$$\text{and linear terms.}$$

$$y = 2(x^2 - 6x) + 2(9) - 2(9) + 10 \qquad \left(\frac{-6}{2}\right)^2 = 9. \text{ Add and subtract } 2(9).$$

$$y = 2(x^2 - 6x + 9) - 8 \qquad\qquad \text{Regroup.}$$

$$y = 2(x - 3)^2 - 8 \qquad\qquad \text{Factor the perfect square trinomial.}$$

$h = 3$ and $k = -8$, so the vertex of the parabola is $(3, -8)$.
The line of symmetry is $x = 3$.

$a = 2$, so the direction of the parabola is the same as the parent function $f(x) = x^2$, but the parabola is stretched vertically.

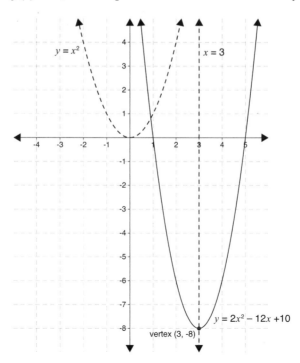

Practice Exercises

Graph each quadratic function using vertex form.

5.28 $y = 2(x-1)^2 - 3$

5.29 $y = x^2 - 4x + 9$

5.30 $y = 3x^2 + 18x - 5$

Comparing Graphing Methods

Three methods of graphing a quadratic function have been presented in this lesson, including making a table of values, using general form, and using vertex form.

While cumbersome, a table of values can be used for an equation in any format. If you forget other methods, making a table of values will always help you identify coordinates on the parabola and draw the graph.

General form and vertex form both yield information about the line of symmetry and the vertex of the parabola. It is helpful to know both general form and vertex form so that you can quickly identify the key facts depending on what form the quadratic function is presented in. You can change from general form to vertex form by completing the square. You can change from vertex form to general form by multiplying and simplifying the function.

	General Form	**Vertex Form**
	$y = ax^2 + bx + c,\ a \neq 0$	$y = a(x-h)^2 + k,\ a \neq 0$
Line of symmetry	$x = \dfrac{-b}{2a}$	$x = h$
Vertex	$x\text{-coordinate} = \dfrac{-b}{2a}$	(h, k)
	Substitute x and solve for y.	
y-intercept	$(0, c)$	

Chapter 6

Polynomials and Polynomial Functions

6.1 Operations with Polynomials

A *monomial* is a real number, a variable, or the product of a real number and one or more variables with whole number exponents. A monomial will not include addition or subtraction. A *binomial* is the sum or difference of two monomials, while a *trinomial* is the sum or difference of three monomials. A *polynomial* is a general term that describes a monomial or the sum or difference of two or more monomials.

Monomial $2, -2xy, x^2, 6x^3yz^2$

Binomial $2x + 3y, 4y - 1, -16x^2y - 3yz^3$

Trinomial $5z - 2x + 3$

Polynomial $2, -2xy, x^2, 6x^3yz^2, 2x + 3y, 4y - 1, -16x^2y - 3yz^3,$
              $5z - 2x + 3, x + y + z + 1$

Multiplying and Dividing Monomials

To simplify a monomial expression means to combine the variables into the simplest form, without parentheses or negative exponents. To simplify a monomial, you must utilize the properties of exponents.

Properties of Exponents

For any real numbers *a* and *b* and positive integers *m* and *n*:

Product of Powers	$a^m \cdot a^n = a^{m+n}$	$x^3 \cdot x^4 = x^7$
Power of a Power	$\left(a^m\right)^n = a^{m \cdot n}$	$\left(x^3\right)^4 = a^{12}$
Power of a Product	$\left(ab\right)^m = a^m b^m$	$\left(2x\right)^3 = 8x^3$
Power of a Quotient	$\left(\dfrac{a}{b}\right)^m = \dfrac{a^m}{b^m}, b \neq 0$	$\left(\dfrac{x}{y}\right)^2 = \dfrac{x^2}{y^2}, y \neq 0$
Quotient of Powers	$\dfrac{a^m}{a^n} = a^{m-n}, a \neq 0$	$\dfrac{x^5}{x^2} = x^3, x \neq 0$
Zero Exponent	$a^0 = 1, a \neq 0$	$x^0 = 1, x \neq 0$
Negative Exponent	$a^{-m} = \dfrac{1}{a^m}, a \neq 0$	$x^{-2} = \dfrac{1}{x^2}, x \neq 0$
	$\dfrac{1}{a^{-n}} = a^n, a \neq 0$	$\dfrac{1}{x^{-3}} = x^3, x \neq 0$

A monomial is fully simplified when:

- No variable appears more than once.
- There are no negative exponents.
- There are no parentheses.
- Fractions are in simplest form.

Example: Simplify each expression.

a. $5m^3(-3m^4)$　　**b.** $\left(-4k^{-5}\right)^2$　　**c.** $\dfrac{24x^4}{3x^3}$　　**d.** $\dfrac{36x^0y^{-2}\left(z^3\right)^2}{4x^{-1}y^{-3}z^{-6}}$

Solution: Use the Properties of Exponents.

a. $5m^3(-3m^4) = -15m^{3+4}$　　Product of Powers.

$\qquad\qquad = -15m^7$　　Add powers.

b. $\left(-4k^{-5}\right)^2 = 16k^{-10}$　　Power of a Power.

$\qquad\qquad = \dfrac{16}{k^{10}}$　　Negative Exponent Property.

c. $\dfrac{24x^4}{3x^3} = 8x^{4-3}$　　Quotient of a Power.

$\qquad\qquad = 8x^1 = 8x$　　Simplify.

d. $\dfrac{36x^0y^{-2}\left(z^3\right)^2}{4x^{-1}y^{-3}z^{-6}} = \dfrac{36(1)y^{-2}z^6}{4x^{-1}y^{-3}z^{-6}}$　　Zero Exponent, Power of a Power.

$\qquad\qquad = \dfrac{9xy^3z^6z^6}{y^2}$　　Negative Exponent Property.

$\qquad\qquad = 9xy^{3-2}z^{6+6}$　　Product and Quotient of Powers.

$\qquad\qquad = 9xyz^{12}$　　Add and subtract powers.

Practice Exercises

Simplify.

6.1　　$2n^{-3}\left(\dfrac{1}{4}n^4\right)$

6.2　　$\left(6a^{-3}b^0\right)^2$

6.3 $\dfrac{5^{-1}y^2}{5y^6}$

6.4 $\dfrac{\left(2x^5y^{-2}\right)^3\left(z^3\right)^0}{8x^2y^{-1}z^2}$

Adding and Subtracting Polynomials

Polynomials can be added and subtracted by adding or subtracting like terms. Recall that like terms are monomials with identical variable expressions.

Like Terms: $2x^2y,\ -5x^2y,\ \dfrac{1}{2}x^2y$

Unlike Terms: $2x^2y,\ -5x^3y,\ \dfrac{1}{2}x^2y^2$

Like terms can be added or subtracted. Unlike terms cannot.

Example: Simplify $(3x^3y+2x^2y-3xy^2)+(-4x^3y+6x^2y-5xy^2)$.

   Solution: Identify and combine like terms.

   $(3x^3y+2x^2y-3xy^2)+(-4x^3y+6x^2y-5xy^2)$

   $=(3x^3y-4x^3y)+(2x^2y+6x^2y)+(-3xy^2-5xy^2)$

   $=-x^3y+8x^2y-8xy^2$

If polynomials are subtracted, be careful to carry the subtraction through to each monomial in the polynomial that is being subtracted.

Example: Simplify $(3x^3y+2x^2y-3xy^2)-(-4x^3y+6x^2y-5xy^2)$.

   Solution: Identify and combine like terms.

   $(3x^3y+2x^2y-3xy^2)-(-4x^3y+6x^2y-5xy^2)$

   $=(3x^3y-(-4x^3y))+(2x^2y-6x^2y)+(-3xy^2-(-5xy^2))$

   $=7x^3y-4x^2y+2xy^2$

Practice Exercises

Simplify.

6.5 $(4x^2y - 3xy - 6xy^2) + (2x^2y - xy + 2xy^2)$

6.6 $(4x^2y - 3xy - 6xy^2) - (2x^2y - xy + 2xy^2)$

Multiplying Polynomials

To multiply a monomial by a polynomial, simply distribute the monomial into the parentheses.

Example: $2x^2y(x^2 + 3x + 2y)$

> **Solution:** Distribute $2x^2y$ into the parentheses.
>
> $2x^2y(x^2 + 3x + 2y) = 2x^4y + 6x^3y + 4x^2y^2$
>
> You can also use a vertical arrangement.
>
> $$\begin{array}{r} x^2 \quad +3x \quad +2y \\ 2x^2y \\ \hline 2x^4y + 6x^3y + 4x^2y^2 \end{array}$$

To multiply two polynomials, multiply each term of one polynomial by each term in the other polynomial. Simplify the answer if possible.

Example: $(3x - 1)(x^2 - 4x + 2)$

> **Solution:** Multiply $3x$ and -1 by $(x^2 - 4x + 2)$.
>
> $(3x - 1)(x^2 - 4x + 2)$
>
> $= 3x(x^2 - 4x + 2) - 1(x^2 - 4x + 2)$ Split into a monomial times a polynomial.
>
> $= 3x^3 - 12x^2 + 6x - x^2 + 4x - 2$ Distribute the monomial.
>
> $= 3x^3 - 13x^2 + 10x - 2$ Combine like terms.

You can also use a vertical arrangement.

$$x^2 - 4x + 2$$
$$\underline{\qquad\qquad 3x - 1}$$
$$-x^2 + 4x - 2$$
$$\underline{3x^3 - 12x^2 + 6x}$$
$$3x^3 - 13x^2 + 10x - 2$$

Practice Exercises

Multiply.

6.7 $-3x^2 y^3 \left(4y^2 - 2y + 6\right)$

6.8 $\left(5x + 2\right)\left(2x^2 + 3x - 1\right)$

6.2 Dividing Polynomials

In Lesson 6.1, you learned how to divide monomials using the rules of exponents.

$$\frac{24x^4}{3x^3} = 8x$$

To divide a polynomial by a monomial, divide each term of the polynomial by the monomial.

Example: Simplify $\dfrac{24x^4 - 6x^3 + 3x^2}{3x^3}$.

Solution: Divide each term of the polynomial by $3x^3$.

$$\frac{24x^4 - 6x^3 + 3x^2}{3x^3}$$

$$= \frac{24x^4}{3x^3} - \frac{6x^3}{3x^3} + \frac{3x^2}{3x^3} \qquad \text{Divide each term by } 3x^3.$$

$$= 8x - 2 + \frac{1}{x} \qquad\qquad \text{Simplify.}$$

Dividing Polynomials with Long Division

To divide a polynomial by another polynomial, you can do long division, similar to the long division you do in arithmetic.

$$\begin{array}{r} 139 \\ 3\overline{)418} \\ \underline{-3} \\ 11 \\ \underline{-9} \\ 28 \\ \underline{-27} \\ 1 \end{array}$$

So, $418 \div 3 = 139\dfrac{1}{3}$

The most challenging part of dividing polynomials by long division is the subtraction step. Be careful to subtract every term.

Example: Divide $4x^2 - 7x - 16$ by $x - 3$.

Solution: Divide the polynomials using long division.

$$\begin{array}{r} 4x \\ x-3\overline{)4x^2\ -7x-16} \\ -\left(4x^2-12x\right) \\ \hline 5x-16 \end{array}$$

Multiply $4x$ by $x - 3$, subtract, and bring down the next term. Remember that $-(4x^2 - 12x)$ is equivalent to $-4x^2 + 12x$.

$$\begin{array}{r} 4x+5 \\ x-3\overline{)4x^2\ -7x-16} \\ -\left(4x^2-12x\right) \\ \hline 5x-16 \\ -\left(5x-15\right) \\ \hline -1 \end{array}$$

Multiply 5 by $x - 3$ and subtract. Remember than $-(5x - 15)$ is equivalent to $-5x + 15$.

$$\begin{array}{r} 4x+5-\dfrac{1}{x-3} \\ x-3\overline{)4x^2\ -7x-16} \\ -\left(4x^2-12x\right) \\ \hline 5x-16 \\ -\left(5x-15\right) \\ \hline -1 \end{array}$$

Place remainder over the divisor.

Therefore, $\dfrac{4x^2-7x-16}{x-3} = 4x+5-\dfrac{1}{x-3}$.

Before you use long division, verify that the terms of the polynomial are in descending order by the highest power of the variable. If there are missing terms, insert these terms with coefficients of 0.

Example: $\left(2x^2 + 4x^4 - 4\right) \div \left(x - 1\right)$

Solution: Arrange the terms of the trinomial in descending order, inserting missing terms with a coefficient of 0. Divide using long division.

$$\left(4x^4 + 0x^3 + 2x^2 + 0x - 4\right) \div \left(x - 1\right)$$

$$\begin{array}{r} 4x^3 + 4x^2 + 6x + 6 + \dfrac{2}{x-1} \\ x-1{\overline{\smash{\big)}\,4x^4 + 0x^3 + 2x^2 + 0x - 4}} \\ -\left(4x^4 - 4x^3\right) \\ \hline 4x^3 + 2x^2 \\ -\left(4x^3 - 4x^2\right) \\ \hline 6x^2 + 0x \\ -\left(6x^2 - 6x\right) \\ \hline 6x - 4 \\ -\left(6x - 6\right) \\ \hline 2 \end{array}$$

There are no x^3 or x terms, so insert these terms with coefficients of 0. Multiply $4x^3$ by $x - 1$, and subtract. Remember that $-(4x^4 - 4x^3)$ is equivalent to $-4x^4 + 4x^3$. Bring down the next term, $2x^2$.

Multiply $4x^2$ by $x - 1$, and subtract. Remember that $-(4x^3 - 4x^2)$ is equivalent to $-4x^3 + 4x^2$. Bring down the next term, $0x$.

Multiply $6x$ by $x - 1$, and subtract. Remember that $-(6x^2 - 6x)$ is equivalent to $-6x^2 + 6x$. Bring down the next term, -4.

Multiply 6 by $x - 1$. Remember that $-(6x - 6)$ is equivalent to $-6x + 6$.

Place the remainder over the divisor.

Therefore, $\left(2x^2 + 4x^4 - 4\right) \div \left(x - 1\right) = 4x^3 + 4x^2 + 6x + 6 + \dfrac{2}{x-1}$.

Practice Exercises

Divide.

6.9 $\left(4y^2 - 6y - 9\right) \div \left(2y - 5\right)$

6.10 $\left(13x - 2x^3 - 15\right) \div \left(x + 3\right)$

Dividing Polynomials with Synthetic Division

As you can see, dividing polynomials using long division takes up a lot of space! Another method that is more efficient is called synthetic division. *Synthetic division* is a method of division that can be used when the divisor is in the form $x - r$, where r is a real number. Observe the simplicity of synthetic division in the next example.

Example: $\left(2x^2 + 4x^4 - 4\right) \div \left(x - 1\right)$

> **Solution:** Arrange the terms of the trinomial in descending order, inserting missing terms with a coefficient of 0. Use synthetic division.
>
> $$\left(4x^4 + 0x^3 + 2x^2 + 0x - 4\right) \div \left(x - 1\right)$$

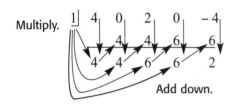

The divisor $x - 1$ is of the form $x - r$ where $r = 1$. Place r in the corner box and place the coefficients of the dividend in a line next to the box. Bring down the first coefficient and multiply r by the first coefficient and place the product underneath the second coefficient. Add down the column. Continue multiplying and adding. The bottom row gives the coefficients of the quotient, with the last element the remainder.

Therefore, just as in the previous example,

$$\left(2x^2 + 4x^4 - 4\right) \div \left(x - 1\right) = 4x^3 + 4x^2 + 6x + 6 + \frac{2}{x - 1}.$$

The previous example illustrates the process of synthetic division.

Steps for Synthetic Division

- Identify r in the divisor $x - r$, and place r in the corner box.

- Write the coefficients for the dividend in descending order, inserting a 0 for any missing term. Bring down the first coefficient.

- Multiply the first coefficient by r and write the product under the second coefficient. Add the product and the second coefficient.

- Repeat the previous step until you have a sum in the last column.

- The numbers on the bottom row are the coefficients of the quotient. The first term has an exponent that is one less than the highest exponent in the dividend, and the exponent decreases with each term. The last number is the remainder.

Synthetic division works only with divisors of the form $x - r$, where the coefficient of x is 1 and r is a real number.

Example: $\left(3x^3 + 2x^2 + 16\right) \div \left(x + 2\right)$

   Solution: Arrange the terms of the trinomial in descending order, inserting missing terms with a coefficient of 0. Identify r. Use synthetic division.

   $\left(3x^3 + 2x^2 + 0x + 16\right) \div \left(x + 2\right);$

   $x + 2 = x - (-2)$, so $r = -2$.

   So $\left(3x^3 + 2x^2 + 16\right) \div \left(x + 2\right) = 3x^2 - 4x + 8$

$$
\begin{array}{r|rrrr}
-2 & 3 & 2 & 0 & 16 \\
   &   & -6 & 8 & -16 \\
\hline
   & 3 & -4 & 8 & 0
\end{array}
$$

Notice in the previous example that if the last number in the bottom row is 0, then there is no remainder and the divisor divides evenly into the dividend.

Synthetic division works only with divisors of the form $x - r$, where the coefficient of x is 1. If the coefficient of x in the divisor is not 1, you must rewrite the divisor.

Example: $\left(6x^3 + 5x^2 - 3x + 2\right) \div \left(2x + 1\right)$

   Solution: Write the division problem as a fraction. Divide the top and bottom of the fraction by the coefficient of x in the divisor. Use synthetic division.

   $$\frac{\left(6x^3 + 5x^2 - 3x + 2\right) \div 2}{\left(2x + 1\right) \div 2} = \frac{3x^3 + \dfrac{5}{2}x^2 - \dfrac{3}{2}x + 1}{x + \dfrac{1}{2}}$$

   The divisor is $x + \dfrac{1}{2}$ or $x - \left(-\dfrac{1}{2}\right)$, so $r = -\dfrac{1}{2}$.

$$-\frac{1}{2}\bigg|\ \begin{array}{cccc} 3 & \dfrac{5}{2} & -\dfrac{3}{2} & 1 \\[2ex] & -\dfrac{3}{2} & -\dfrac{1}{2} & 1 \\[2ex] \hline 3 & 1 & -2 & 2 \end{array}$$

So, $\left(6x^3 + 5x^2 - 3x + 2\right) \div \left(2x + 1\right) = 3x^2 + x - 2 + \dfrac{2}{x + \dfrac{1}{2}}$

It is not common practice to leave a fraction in the denominator in a solution. Rewrite the remainder by multiplying the top and bottom of the whole fraction by a number that will eliminate the fraction in the denominator.

$$\frac{2}{x + \dfrac{1}{2}} \cdot \frac{2}{2} = \frac{2 \cdot 2}{\left(x + \dfrac{1}{2}\right) \cdot 2} = \frac{4}{2x + 1}$$

Therefore, $\left(6x^3 + 5x^2 - 3x + 2\right) \div \left(2x + 1\right) = 3x^2 + x - 2 + \dfrac{4}{2x + 1}$.

Practice Exercises

Use synthetic division.

6.11 $\left(3x^3 + 2x^2 - 6x - 20\right) \div \left(x - 2\right)$

6.12 $\left(x^3 + 6x^2 + 3x - 8\right) \div \left(x + 5\right)$

6.13 $\left(6x^4 + 7x^3 + 2x - 1\right) \div \left(3x - 1\right)$

6.3 Solving Polynomial Equations by Factoring

In Lesson 5.2, you learned how to solve a quadratic equation by factoring. You used different factoring methods including factoring out common factors and using factor patterns such as the difference of two perfect squares.

Factoring Polynomials

The factor patterns you learned in Lesson 5.2 and additional factor patterns can be useful in factoring polynomial expressions.

Cubic Factor Patterns

- Sum of two cubes

    General Form　　$a^3 + b^3 = (a+b)(a^2 - ab + b^2)$

    Example　　$x^3 + 8 = x^3 + 2^3 = (x+2)(x^2 - 2x + 4)$

- Difference of two cubes

    General Form　　$a^3 - b^3 = (a-b)(a^2 + ab + b^2)$

    Example　　$x^3 - 27 = x^3 - 3^3 = (x-3)(x^2 + 3x + 9)$

- Quadratic Form

    General Form　　$a(u)^2 + b(u) + c$ where u is a variable expression

    　　　　　　　　Find factors of ac whose sum is b and factor as a trinomial.

    Example　　$x^4 + 10x^2 + 21 = (x^2)^2 + 10(x^2) + 21$

    　　　　　　Let $u = x^2$, then $u^2 + 10u + 21 = (u+3)(u+7)$.

    　　　　　　Let $x^2 = u$, then $(u+3)(u+7) = (x^2+3)(x^2+7)$.

    　　　　　　So $x^4 + 10x^2 + 21 = (x^2+3)(x^2+7)$.

Always factor out common factors before looking for factor patterns.

Example: Factor $40x^4 + 5x$.

　　Solution: Factor out the common factor. Use the sum of cubes factor pattern.

　　$40x^4 + 5x$

　　$= 5x(8x^3 + 1)$　　　　　　　　Factor $5x$ from both terms.

　　$= 5x\left((2x)^3 + 1^3\right)$　　　　　　Rewrite terms as perfect cubes.

　　$= 5x(2x+1)(4x^2 - 2x + 1)$　Use the sum of two cubes factor pattern.

Notice in the previous example that $4x^2 - 2x + 1$ cannot be further factored since no factors of $ac = 4$ add to be $-2 = b$.

The *leading coefficient* of a polynomial is the coefficient of the variable with the highest exponent. It is often helpful to factor a polynomial so that the leading coefficient is positive.

Example: Factor $-12x^6 - 18x^3 + 12$.

> **Solution:** Factor out -6 so that the leading coefficient is positive. Use the quadratic form factor pattern.
>
> $-12x^6 - 18x^3 + 12$
>
> $= -6(2x^6 + 3x^3 - 2)$　　　　Factor -6 from all terms.
>
> $= -6\left[2\left(x^3\right)^2 + 3\left(x^3\right) - 2\right]$　　Write the polynomial in quadratic form.
>
> $= -6\left(2u^2 + 3u - 2\right)$　　　Substitute $u = x^3$.
>
> $= -6\left(2u^2 + 4u - 1u - 2\right)$　　$ac = -4$, $4 + (-1) = 3 = b$.
>
> $= -6\left[2u(u+2) - (u+2)\right]$　Factor by grouping.
>
> $= -6(2u - 1)(u + 2)$　　　Factor by grouping.
>
> $= -6(2x^3 - 1)(x^3 + 2)$　　Substitute $x^3 = u$.

Notice in the previous example that neither $2x^3 - 1$ nor $x^3 + 2$ are the sum or difference of perfect cubes, so they cannot be factored further.

To completely factor a polynomial, factor out any common factors first, and then use other factor techniques. The table on the next page shows which techniques are effective based on the number of terms in the polynomial.

Number of Terms	Factor Technique	Examples
Any	Factor out common factors	$ab - b^2 = b(a - b)$ $3x^2 - 6x = 3x(x - 2)$
2	Difference of perfect squares	$a^2 - b^2 = (a + b)(a - b)$ $4x^2 - 9 = (2x + 3)(2x - 3)$
	Sum of perfect cubes	$a^3 + b^3 = (a + b)(a^2 - ab + b^2)$ $x^3 + 8 = x^3 + 2^3 = (x + 2)(x^2 - 2x + 4)$
	Difference of perfect cubes	$a^3 - b^3 = (a - b)(a^2 + ab + b^2)$ $x^3 - 27 = x^3 - 3^3 = (x - 3)(x^2 + 3x + 9)$
3	Perfect square trinomial	$a^2 + 2ab + b^2 = (a + b)^2$ $x^2 + 6x + 9 = x^2 + 2(3x) + 3^2 = (x + 3)^2$ $a^2 - 2ab + b^2 = (a - b)^2$ $x^2 - 8x + 16 = x^2 - 2(4x) + 4^2 = (x - 4)^2$
	General trinomial	$ax^2 + bx + c$ Find factors of ac whose sum is b and factor by grouping. $x^2 - 5x + 6$; $ac = 6$, $-3 + -2 = -5 = b$ $x^2 - 5x + 6 = x^2 - 3x - 2x + 6$ $= x(x - 3) - 2(x - 3)$ $= (x - 3)(x - 2)$
	Quadratic form	$a(u)^2 + bu + c$ where u is an expression. Factor as a general trinomial. $x^4 + 10x^2 + 21 = (x^2)^2 + 10(x^2) + 21$ Let $u = x^2$, then $u^2 + 10u + 21 = (u+3)(u+7)$. Let $x^2 = u$, then $(u+3)(u+7) = (x^2+3)(x^2+7)$. So $x^4 + 10x^2 + 21 = (x^2 + 3)(x^2 + 7)$.

4 or more   Grouping

$$ax + bx + ay + by = x(a + b) + y(a + b)$$
$$= (x + y)(a + b)$$

$$6x^2 - 2x + 3xy - y = 2x(3x - 1) + y(3x - 1)$$
$$= (2x + y)(3x - 1)$$

A polynomial that cannot be factored is *prime*. A polynomial that is completely factored will be the product of prime factors.

Example: Factor completely $2x^4 + 6x^3 - 8x^2 - 24x$.

> **Solution:** Factor out the common factor and then factor by grouping. Continue to factor until all of the factors are prime.

$$2x^4 + 6x^3 - 8x^2 - 24x$$

$$= 2x(x^3 + 3x^2 - 4x - 12) \quad \text{Factor out the common factor.}$$

$$= 2x\left[x^2(x+3) - 4(x+3) \right] \quad \text{Factor by grouping.}$$

$$= 2x(x^2 - 4)(x + 3) \quad \text{Factor by grouping.}$$

$$= 2x(x + 2)(x - 2)(x + 3) \quad \text{Use the difference of perfect squares factor pattern.}$$

Example: Factor completely $x^5 + 2x^4y + x^3y^2 + x^2y^3 + 2xy^4 + y^5$.

Solution: There are no common factors. Factor by grouping. Continue to factor until all of the factors are prime.

$$x^5 + 2x^4y + x^3y^2 + x^2y^3 + 2xy^4 + y^5$$

$$= (x^5 + 2x^4y + x^3y^2) + (x^2y^3 + 2xy^4 + y^5) \quad \text{Group.}$$

$$= x^3(x^2 + 2xy + y^2) + y^3(x^2 + 2xy + y^2) \quad \text{Factor by grouping.}$$

$$= (x^3 + y^3)(x^2 + 2xy + y^2) \quad \text{Factor by grouping.}$$

$$= (x^3 + y^3)(x + y)^2 \quad \text{Use perfect square trinomial pattern.}$$

$$= (x + y)(x^2 - xy + y^2)(x + y)^2 \quad \text{Use sum of perfect cubes pattern.}$$

Practice Exercises

Factor completely.

6.14　$192x^5 - 81x^2$

6.15　$x^4 - 81$

6.16　$4xy + 8xz + 6y + 12z$

Solving Polynomial Equations

A polynomial expression in one variable has the form

$a_n x^n + a_{n-1}x^{n-1} + ... + a_2 x^2 + a_1 x + a_0$, where $a_n \neq 0$, $a_{n-1}, ..., a_2, a_1, a_0$ are real numbers, and n is a whole number. The *degree* of the polynomial is the value of the greatest exponent. The *leading coefficient* is the coefficient of the variable with the greatest exponent. A polynomial is in *standard form* if the expression is written so that the exponents are in descending order.

Degree	Polynomial	Example	Leading Coefficient
0	Constant	12	12
1	Linear	$2x + 1$	2
2	Quadratic	$x^2 - 2x + 4$	1
3	Cubic	$5x^3 - 2x^2 + x - 3$	5
n	General	$a_n x^n + a_{n-1}x^{n-1} + ... + a_2 x^2 + a_1 x + a_0$	a_n

A polynomial equation is a polynomial expression that is set equal to a value. You solved linear equations in one variable in Chapter 1. In Chapter 5 you learned techniques for solving quadratic equations in one variable.

In Lesson 5.2 you applied the Zero Product Property to solve quadratic equations. Recall that the Zero Product Property states that if a product equals zero, then at least one of the factors is equal to zero. You can use the Zero Product Property to solve polynomial equations of higher degree.

Example: Solve. $2x^5 - 26x^3 + 72x = 0$

Solution: Factor out the common factor. Use the quadratic factor pattern and the Zero Product Property.

$$2x^5 - 26x^3 + 72x = 0$$

$$2x(x^4 - 13x^2 + 36) = 0 \qquad \text{Factor out the common factor.}$$

$$2x\left((x^2)^2 - 13(x^2) + 36\right) = 0 \qquad \text{Use quadratic form.}$$

$$2x(u^2 - 13u + 36) = 0 \qquad \text{Substitute } u = x^2$$

$$2x(u^2 - 9u - 4u + 36) = 0 \qquad ac = 36,\ -9 - 4 = -13 = b$$

$$2x\left[u(u - 9) - 4(u - 9)\right] = 0 \qquad \text{Factor by grouping.}$$

$$2x\left[(u - 9)(u - 4)\right] = 0 \qquad \text{Factor by grouping.}$$

$$2x\left[(x^2 - 9)(x^2 - 4)\right] = 0 \qquad \text{Substitute } x^2 = u.$$

$2x = 0$ or $x^2 - 9 = 0$ or $x^2 - 4 = 0$ Use the Zero Product Property.

$\quad x = 0 \quad$ or $\quad x^2 = 9 \quad$ or $\quad x^2 = 4 \quad$ Simplify.

$\qquad\qquad\qquad x = \pm 3 \quad$ or $\quad x = \pm 2 \quad$ Use the Square Root Property.

Check:

$x = 0:\ 2(0)^5 - 26(0)^3 + 72(0) = 0$ ✓

$x = -3:\ 2(-3)^5 - 26(-3)^3 + 72(-3) = -486 + 702 - 216 = 0$ ✓

$x = 3:\ 2(3)^5 - 26(3)^3 + 72(3) = 486 - 702 + 216 = 0$ ✓

$x = -2:\ 2(-2)^5 - 26(-2)^3 + 72(-2) = -64 + 208 - 144 = 0$ ✓

$x = 2:\ 2(2)^5 - 26(2)^3 + 72(2) = 64 - 208 + 144 = 0$ ✓

The solutions to the polynomial equation are $-3, -2, 0, 2, 3$. Sometimes polynomial equations will have complex solutions.

Example: Solve. $x^3 + 1 = 0$

Solution: Factor the polynomial with the sum of perfect cubes factor pattern. Use the Zero Product Property. Solve the quadratic equation by completing the square or using the quadratic formula.

$$x^3 + 1 = 0$$

$$(x + 1)(x^2 - x + 1) = 0 \qquad \text{Use the sum of perfect cubes pattern.}$$

$$x + 1 = 0 \text{ or } x^2 - x + 1 = 0 \qquad \text{Use the Zero Product Property.}$$

$$x = -1 \text{ or } x = \frac{-(-1) \pm \sqrt{(-1)^2 - 4(1)(1)}}{2(1)} \qquad \text{Solve for } x.$$

$$x = \frac{-(-1) \pm \sqrt{-3}}{2(1)} \qquad \text{Simplify.}$$

$$x = \frac{1 \pm i\sqrt{3}}{2} \qquad \text{Simplify.}$$

Check:

$$x = -1: \ (-1)^3 + 1 = -1 + 1 = 0 \ \checkmark$$

$$x = \frac{1 + i\sqrt{3}}{2}: \qquad \left(\frac{1 + i\sqrt{3}}{2}\right)^3 + 1 = 0$$

$$\left(\frac{1 + i\sqrt{3}}{2}\right)\left(\frac{1 + i\sqrt{3}}{2}\right)\left(\frac{1 + i\sqrt{3}}{2}\right) + 1 = 0 \qquad \text{Definition of exponent.}$$

$$\left(\frac{1 + 2i\sqrt{3} + \left(i\sqrt{3}\right)^2}{4}\right)\left(\frac{1 + i\sqrt{3}}{2}\right) + 1 = 0 \qquad \text{Multiply.}$$

$$\left(\frac{-2 + 2i\sqrt{3}}{4}\right)\left(\frac{1 + i\sqrt{3}}{2}\right) + 1 = 0 \qquad \text{Simplify.}$$

$$\left(\frac{-1 + i\sqrt{3}}{2}\right)\left(\frac{1 + i\sqrt{3}}{2}\right) + 1 = 0 \qquad \text{Simplify.}$$

$$\left(\frac{-1-i\sqrt{3}+i\sqrt{3}+\left(i\sqrt{3}\right)^2}{4}\right)+1=0 \qquad \text{Multiply.}$$

$$\frac{-4}{4}+1=0 \checkmark \qquad \text{Simplify.}$$

$x=\dfrac{1-i\sqrt{3}}{2}$: The check is left to you.

The solutions are -1, $\dfrac{1\pm\sqrt{3}}{2}$

In the previous example, you could have completed the square to solve the equation $x^2-x+1=0$. Note the sequence of steps:

$x^2-x=-1$

$x^2-x+\dfrac{1}{4}=-1+\dfrac{1}{4}$ $\qquad \left(-1\cdot\dfrac{1}{2}\right)^2=\dfrac{1}{4}$. Add $\dfrac{1}{4}$ to both sides.

$\left(x-\dfrac{1}{2}\right)^2=-\dfrac{3}{4}$ $\qquad\qquad$ Factor the perfect square trinomial.

$x-\dfrac{1}{2}=\pm\sqrt{-\dfrac{3}{4}}$ $\qquad\qquad$ Use the Square Root Property.

$x=\dfrac{1}{2}\pm\sqrt{-\dfrac{3}{4}}=\dfrac{1}{2}\pm\dfrac{i\sqrt{3}}{2}=\dfrac{1\pm i\sqrt{3}}{2}$ $\qquad$ Simplify.

Practice Exercises

Solve.

6.17 $\quad x^4+6x^2-16=0$

6.18 $\quad x^3-125=0$

6.4 Polynomial Functions

A *polynomial function* has the form $P(x)=a_nx^n+a_{n-1}x^{n-1}+...+a_2x^2+a_1x+a_0$, where $a_n\neq 0$, $a_{n-1},..., a_2, a_1, a_0$ are real numbers, and n is a whole number. Just like other functions, a polynomial function is evaluated by substituting a value for x.

Example: Evaluate $P(-2)$ if $P(x) = 4x^3 - 2x^2 + 7x - 9$.

Solution: Substitute -2 for x in the function and simplify.

$P(x) = 4x^3 - 2x^2 + 7x - 9$

$P(-2) = 4(-2)^3 - 2(-2)^2 + 7(-2) - 9$

$P(-2) = 4(-8) - 2(4) - 14 - 9$

$P(-2) = -32 - 8 - 14 - 9$

$P(-2) = -63$

Functions can also be evaluated for algebraic expressions.

Example: Evaluate $P(x + h)$ if $P(x) = 4x^3 - 2x^2 + 7x - 9$.

Solution: Substitute $x + h$ for x in the function and simplify.

$P(x) = 4x^3 - 2x^2 + 7x - 9$

$P(x + h) = 4(x + h)^3 - 2(x + h)^2 + 7(x + h) - 9$

$P(x + h) = 4(x + h)^2(x + h) - 2(x + h)^2 + 7(x + h) - 9$

$P(x + h) = 4(x^2 + 2xh + h^2)(x + h) - 2(x^2 + 2xh + h^2) + 7(x + h) - 9$

$P(x + h) = 4(x^3 + 3x^2h + 3xh^2 + h^3) - 2(x^2 + 2xh + h^2) + 7(x + h) - 9$

$P(x + h) = 4x^3 + 12x^2h + 12xh^2 + 4h^3 - 2x^2 - 4xh - 2h^2 + 7x + 7h - 9$

Practice Exercises

$P(x) = x^2 - 3x + 4$

6.19   Find $P(4)$.

6.20   Find $P(x - h)$.

6.5 Graphing Polynomial Functions

The most basic polynomial functions are called *power functions* and have the form $f(x) = x^n$. These functions are also the parent functions for polynomial functions. The graphs of the power functions have two basic shapes, depending on whether the degree of the function is even or odd.

$f(x) = x^n$, n is even $\qquad\qquad$ $f(x) = x^n$, n is odd

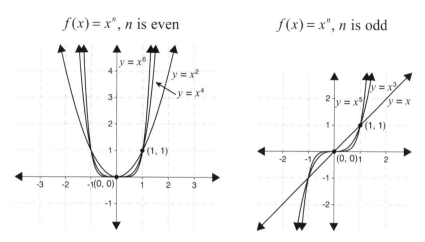

Notice that all of the graphs of the power functions pass through the common points $(0, 0)$, and $(1, 1)$.

Recall from Chapter 4 that the opposite of the parent function, $f(x) = -x^2$, reflects the graph of $f(x) = x^2$ over the x-axis. This is true for all of the power functions as well. In fact, for any polynomial function, $P(x) = a_n x^n + a_{n-1} x^{n-1} + ... + a_2 x^2 + a_1 x + a_0$, if the leading coefficient a_n is negative, the graph will be reflected over the x-axis from the parent function.

$f(x) = -x^n$, n is even $\qquad\qquad$ $f(x) = -x^n$, n is odd

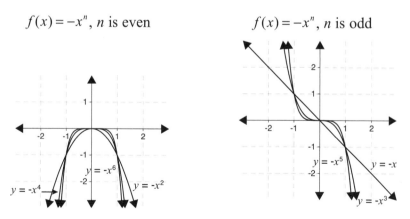

Notice the behavior of the graphs at the left and right ends, as x approaches negative infinity and positive infinity. The end behavior of the graph of a polynomial function is determined by the degree of the function and the sign of the leading coefficient. Compare the following table to the graphs of the power functions.

Degree	Ends of Graph	Leading Coefficient	Left End	Right End
Even	Same direction	Positive	Up	Up
Even	Same direction	Negative	Down	Down
Odd	Opposite directions	Positive	Down	Up
Odd	Opposite directions	Negative	Up	Down

While transformations of the power functions work like you have learned for the parent function $f(x) = x^2$, it can be difficult to factor polynomial functions of higher degree into the proper form to graph by transformations. Other techniques are typically used for functions of degree three and higher.

If the polynomial can be factored, you can locate where the graph intersects the x-axis by finding the values of x that make $f(x) = 0$. These values are called the *zeros* of the function. The degree of the function is equal to the maximum number of zeros the function can have.

Example: Graph $f(x) = x^4 - 5x^2 + 4$.

Solution: Since the degree of the function is 4, it can have at most four zeros. Set the function equal to 0. Solve the equation by factoring to find the zeros of the function. Graph the zeros. Choose x values between the zeros and on the left and right end of the graph and evaluate the function at these points. Graph the new points and make a sketch of the graph.

$f(x) = x^4 - 5x^2 + 4$

$$x^4 - 5x^2 + 4 = 0 \quad \text{Set the function equal to 0.}$$

$$\left(x^2\right)^2 - 5\left(x^2\right) + 4 = 0 \quad \text{Use the quadratic form.}$$

$$u^2 - 5u + 4 = 0 \quad \text{Substitute } u = x^2.$$

$$(u - 4)(u - 1) = 0 \quad \text{Factor. } ac = 4; -1 + -4 = -5 = b.$$

$$(x^2 - 4)(x^2 - 1) = 0 \quad \text{Substitute } x^2 = u.$$

$$(x + 2)(x - 2)(x + 1)(x - 1) = 0 \quad \text{Use the difference of perfect squares factor pattern.}$$

$$x = -2, 2, -1, 1 \quad \text{Use the Zero Product Property and simplify.}$$

The zeros of the function are –2, 2, –1, and 1. So the graph intersects the x-axis at –2, 2, –1, and 1. In other words, the points (–2, 0), (2, 0), (–1, 0) and (1, 0) are on the graph of the function.

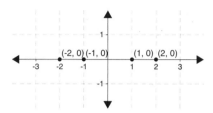

x	$x^4 - 5x^2 + 4$	f(x)	
–3	$(-3)^4 - 5(-3)^2 + 4$	40	Choose x values between the zeros and at the ends of the graph.
–1.5	$(-1.5)^4 - 5(-1.5)^2 + 4$	–2.1875	
0	$(0)^4 - 5(0)^2 + 4$	4	Evaluate the function for each of the values.
1.5	$(1.5)^4 - 5(1.5)^2 + 4$	–2.1875	
3	$(3)^4 - 5(3)^2 + 4$	40	

Plot the points and draw a smooth curve.

Notice in the previous example how the function is related to the parent function $f(x) = x^4$. Since the degree of the function is even and the leading coefficient is positive, the left and right ends of the graph both go up just like the parent function.

In the previous example, factoring enabled you to locate the zeros. However, many polynomials cannot be factored. Another method for graphing is to use a table of values.

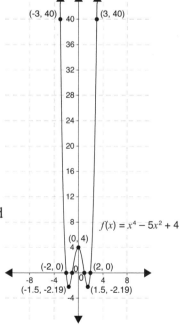

Example: Graph $P(x) = -3x^3 + 4x - 1$.

 Solution: The polynomial is not readily factorable. Since the degree of the polynomial is odd, the ends of the graph will go in opposite directions. Since the leading coefficient is negative, the left end of the graph will go up and the right end of the graph will go down. Choose various values for x and evaluate $f(x)$. Since the degree of the function is 3, it can have at most three zeros.

x	$-3x^3 + 4x - 1$	f(x)	x	$-3x^3 + 4x - 1$	f(x)
–2	$-3(-2)^3 + 4(-2) - 1$	15	0.5	$-3(0.5)^3 + 4(0.5) - 1$	0.625
–1.5	$-3(-1.5)^3 + 4(-1.5) - 1$	3.125	1	$-3(1)^3 + 4(1) - 1$	0
–1	$-3(-1)^3 + 4(-1) - 1$	–2	1.5	$-3(1.5)^3 + 4(1.5) - 1$	–5.125
–0.5	$-3(-0.5)^3 + 4(-0.5) - 1$	–2.625	2	$-3(2)^3 + 4(2) - 1$	–17
0	$-3(0)^3 + 4(0) - 1$	–1			

Graph the points and draw a smooth curve.

In the previous example you can see from the graph that there are zeros that lie between –1 and –1.5 and between 0 and 0.5 because the graph crosses the x-axis between these values of x. Notice the changes in value in $f(x)$ for those x values:

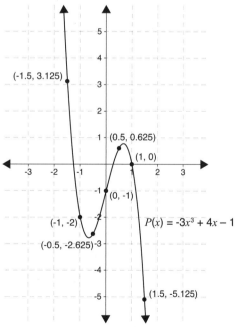

x	f(x)
–1.5	3.125
–1	–2
0	–1
0.5	0.625

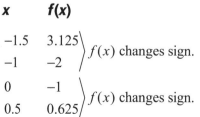

These examples of a change in sign illustrate the Location Theorem.

Location Theorem

If $y = f(x)$ is a polynomial function and $f(a) < 0$ and $f(b) > 0$ for real numbers a and b, then there is at least one real zero between a and b.

Graphing Polynomial Functions

- The behavior of the graph at the left and right ends is determined by the degree and the leading coefficient.

- The degree of the polynomial is equal to the maximum number of zeros the function can have.

- The zeros of the function are where $f(x) = 0$.

- If there is a sign change between $f(a)$ and $f(b)$, then there is at least one zero between a and b.

Example: Graph $f(x) = 2x^3 + 3x^2 - 3x - 2$.

Solution: The degree of the function is 3 and the leading coefficient is positive, so the left end of the graph will go down and the right end of the graph will go up. The polynomial is not readily factorable, so make a table of values. The maximum number of zeros is three since the degree of the function is 3.

x	$2x^3 + 3x^2 - 3x - 2$	$f(x)$
–4	$2(-4)^3 + 3(-4)^2 - 3(-4) - 2$	–70
–3	$2(-3)^3 + 3(-3)^2 - 3(-3) - 2$	–20
–2	$2(-2)^3 + 3(-2)^2 - 3(-2) - 2$	0 → Zero
–1	$2(-1)^3 + 3(-1)^2 - 3(-1) - 2$	2
0	$2(0)^3 + 3(0)^2 - 3(0) - 2$	–2 → $f(x)$ changes sign between $x = -1$ and 0
1	$2(1)^3 + 3(1)^2 - 3(1) - 2$	0 → Zero
2	$2(2)^3 + 3(2)^2 - 3(2) - 2$	20
3	$2(3)^3 + 3(3)^2 - 3(3) - 2$	70

There are zeros at –2, 1, and between –1 and 0. The values of $f(x)$ are decreasing to negative infinity on the left and to positive infinity on the right. Using the values from the table and the end behavior of the graph, make a sketch of the graph.

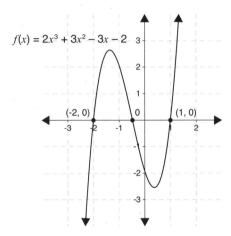

$f(x) = 2x^3 + 3x^2 - 3x - 2$

Practice Exercises

Graph each function.

6.21 $f(x) = x^4 + 2x^2 - 3$

6.22 $f(x) = 4x^3 + 4x^2 - 11x - 6$

6.6 Remainder and Factor Theorems

The last example of Lesson 6.5 illustrates that there are polynomials that have integral zeros and yet the polynomial does not appear readily factorable. The Remainder and Factor Theorems will give you additional tools to identify factors of polynomials.

Remainder Theorem

Recall that when you divide using long division or synthetic division, there is often a remainder. For example, the synthetic division on the next page shows that $(4x^4 + 2x^2 - 4) \div (x - 1) = 4x^3 + 4x^2 + 6x + 6$ with a remainder of 2.

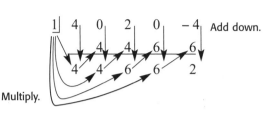

The Remainder Theorem connects the value of the remainder when you divide $P(x)$ by $x - r$ and the value of $P(r)$.

Remainder Theorem

If a polynomial $P(x)$ is divided by $(x - r)$, then the remainder is equal to the value of $P(r)$ and the value of $P(r)$ is equal to the remainder when $P(x)$ is divided by $(x - r)$.

Example: Find the value of $P(-2)$ if $P(x) = -2x^6 + 3x^3 - x + 4$.

Solution: The Remainder Theorem provides two ways to find the value of $P(-2)$. One method is to substitute -2 for x in $P(x)$. Another method is to divide $P(x)$ by $x - (-2)$.

Method 1: Substitute -2 for x in $P(x)$ and simplify.

$$P(x) = -2x^6 + 3x^3 - x + 4$$

$$P(-2) = -2(-2)^6 + 3(-2)^3 - (-2) + 4$$

$$P(-2) = -2(64) + 3(-8) - (-2) + 4$$

$$P(-2) = -128 - 24 + 2 + 4$$

$$P(-2) = -146$$

Method 2: Divide $P(x)$ by $x - (-2)$.

Recall the synthetic division method taught in Lesson 6.2. Remember to insert missing terms with a coefficient of 0.

$$P(x) = -2x^6 + 0x^5 + 0x^4 + 3x^3 + 0x^2 - x + 4$$

$$x - r = x - (-2); \; r = -2$$

$$
\begin{array}{r|rrrrrrr}
-2 & -2 & 0 & 0 & 3 & 0 & -1 & 4 \\
   &   & 4 & -8 & 16 & -38 & 76 & -150 \\
\hline
   & -2 & 4 & -8 & 19 & -38 & 75 & -146
\end{array}
$$

So, $P(-2) = -146$.

The Remainder Theorem might seem like just a cute little trick, but there are broader implications.

Example: Find the remainder if $P(x) = 2x^3 + 3x^2 - 3x - 2$ is divided by $x - 1$.

    Solution: The Remainder Theorem provides two ways to find the remainder of $P(x) \div (x - 1)$. One method is to substitute 1 for x in $P(x)$ and find $P(1)$. Another method is to divide $P(x)$ by $x - 1$ using synthetic division.

    Method 1: Substitute 1 for x in $P(x)$ and find $P(1)$.

$$P(x) = 2x^3 + 3x^2 - 3x - 2$$

$$P(1) = 2(1)^3 + 3(1)^2 - 3(1) - 2$$

$$P(1) = 2 + 3 - 3 - 2$$

$$P(1) = 0$$

    Method 2: Divide $P(x)$ by $x - 1$ using synthetic division.

$$P(x) = 2x^3 + 3x^2 - 3x - 2$$

$$x - r = x - 1;\ r = 1$$

$$\underline{1|}\ \begin{array}{rrrr} 2 & 3 & -3 & -2 \\ & 2 & 5 & 2 \\ \hline 2 & 5 & 2 & 0 \end{array}$$ The remainder is 0.

In the previous example, $(2x^3 + 3x^2 - 3x - 2) \div (x - 1) = 0$ means that $x - 1$ is a factor of $P(x)$ and that $P(1) = 0$. If $P(1) = 0$, then 1 is a zero of $P(x)$. Notice that this is the same polynomial function from the last example in Lesson 6.5 when you identified the zeros by making a table of values.

Practice Exercises

$$P(x) = 3x^4 - 2x^3 + 4x - 2$$

6.23   Find $P(-2)$.

6.24   Find the remainder of $P(x) \div (x + 2)$

Factor Theorem

The Factor Theorem is a special case of the Remainder Theorem, when the remainder is zero.

> **Factor Theorem**
>
> If $P(x)$ is a polynomial, then $P(r) = 0$ if and only if $x - r$ is a factor of $P(x)$.

You can use the Factor Theorem to factor polynomials and identify zeros of a function.

Example: Is $x - 3$ a factor of $P(x) = x^3 + 2x^2 - 11x - 12$? If so, find the remaining factors and the zeros of $P(x)$.

Solution: Divide $x^3 + 2x^2 - 11x - 12$ by $x - 3$ using synthetic division and see if the remainder is 0. If so, determine whether the resulting polynomial is factorable.

$$\underline{3|}\ \begin{array}{rrrr} 1 & 2 & -11 & -12 \\ & 3 & 15 & 12 \\ \hline 1 & 5 & 4 & 0 \end{array}$$

Since the remainder is 0, $x - 3$ is a factor of $x^3 + 2x^2 - 11x - 12$, and by the Factor Theorem, $P(3) = 0$ and 3 is a zero of $P(x)$.

The synthetic division shows that
$(x^3 + 2x^2 - 11x - 12) \div (x - 3) = x^2 + 5x + 4.$

Or, $x^3 + 2x^2 - 11x - 12 = (x - 3)(x^2 + 5x + 4)$.

$x^2 + 5x + 4 = (x + 4)(x + 1)$, so

$x^3 + 2x^2 - 11x - 12 = (x - 3)(x + 4)(x + 1)$.

The factors of $P(x)$ are $(x - 3)$, $(x + 4)$ and $(x + 1)$.

By the Factor Theorem, $P(3) = 0$, $P(-4) = 0$ and $P(-1) = 0$.

If $P(r) = 0$, then r is a zero of $P(x)$.

So the zeros of $P(x)$ are 3, –4, and –1.

Practice Exercises

Is $x - 1$ a factor of $P(x)$? If so, find the remaining factors of $P(x)$ and the zeros of $P(x)$.

6.25 $P(x) = x^3 - x^2 - 4x + 4$

6.26 $P(x) = x^3 - 5x^2 - 8x + 12$

6.7 Finding All Factors, Zeros, and Solutions

In this lesson, we'll pull together concepts and extend them in order to analyze polynomials, polynomial equations, and functions as completely as possible.

Relationships Among Factors, Zeros, and Solutions

If $P(x)$ is a polynomial function such that $P(x) = a_n x^n + a_{n-1} x^{n-1} + ... + a_2 x^2 + a_1 x + a_0$, then all of the bulleted statements below are equivalent.

- r is a zero of $P(x)$.
- $P(r) = 0$.
- r is a solution to the polynomial equation
  $a_n x^n + a_{n-1} x^{n-1} + ... + a_2 x^2 + a_1 x + a_0 = 0$.
- $x - r$ is a factor of the polynomial $a_n x^n + a_{n-1} x^{n-1} + ... + a_2 x^2 + a_1 x + a_0$.
- If r is real, then r is an x-intercept of the graph of $P(x)$, so $(r, 0)$ is on the graph of $P(x)$.

Example: The zeros of $P(x) = 2x^3 + x^2 - 5x + 2$ are -2, 1, and $\dfrac{1}{2}$.

Based on the relationships among factors, zeros, and solutions, what other statements must be true?

Solution:

For each of the zeros, $P(r) = 0$, so $P(-2) = 0$, $P(1) = 0$ and $P\left(\dfrac{1}{2}\right) = 0$.

-2, 1, $\dfrac{1}{2}$ are each solutions to the polynomial equation

$2x^3 + x^2 - 5x + 2 = 0$.

$(x + 2)$, $(x - 1)$ and $\left(x - \dfrac{1}{2}\right)$ are factors of the polynomial

$2x^3 + x^2 - 5x + 2$.

-2, 1, $\dfrac{1}{2}$ are each x-intercepts of the graph of $P(x) = 2x^3 + x^2 - 5x + 2$,

and $(-2, 0)$, $(1, 0)$, and $\left(\dfrac{1}{2}, 0\right)$ all lie on the graph of $P(x)$.

The Rational Zero Theorem

The previous lesson showed you how to verify that a binomial is a factor of a polynomial or the zero of the polynomial function. The Rational Zero Theorem can help you choose the values of r to test.

Rational Zero Theorem

If a polynomial function with integral coefficients has a rational zero, then the zero will have the form $\dfrac{p}{q}$ (in simplest form), with p a factor of the constant term and q a factor of the leading coefficient.

Corollary: If a polynomial function with integral coefficients and a leading coefficient of 1 has a rational zero, then the zero will be a factor of the constant term.

Example: Identify the possible rational zeros for the function

$$P(x) = 2x^3 + 7x^2 - 5x - 4.$$

Solution: Identify the factors p of the constant term and the factors of the leading coefficient q. The Rational Zero Theorem says that any rational zeros will have the form $\dfrac{p}{q}$.

p: factors of -4: ± 1, ± 2, ± 4

q: factors of 2: ± 1, ± 2

$\dfrac{p}{q}$: ± 1, $\pm \dfrac{1}{2}$, ± 2, ± 4

As you can see, a function can have quite a few values to check. Recall that you can check $\dfrac{p}{q}$ by substituting it into the function to see if

$P\left(\dfrac{p}{q}\right) = 0$, or you can use synthetic division to see if $\left(x - \dfrac{p}{q}\right)$ is a factor

of $P(x)$. It is most efficient to use synthetic division so that you can easily write the function in factored form. You can use a compressed form of synthetic division notation to simplify your work.

Using the possible zeros from the previous example, you can check the values -1 and 1:

Traditional Synthetic Division

$$
\begin{array}{r|rrrr}
-1 & 2 & 7 & -5 & -4 \\
& & -2 & -5 & 10 \\
\hline
& 2 & 5 & -10 & 6
\end{array}
$$

$$
\begin{array}{r|rrrr}
1 & 2 & 7 & -5 & -4 \\
& & 2 & 9 & 4 \\
\hline
& 2 & 9 & 4 & 0
\end{array}
$$

Compressed Form

$\dfrac{p}{q}$	2	7	-5	-4
-1	2	5	-10	6
1	2	9	4	0

1 is a zero and $x - 1$ is a factor of $P(x)$.

Notice that the middle line of the traditional synthetic division notation is eliminated in the compressed form.

Example: Find the rational zeros of $P(x) = x^3 - 3x^2 - 6x + 8$.

Solution: Identify and test the possible rational zeros of the function.

The Corollary to the Rational Zero Theorem says that since the leading coefficient is 1, if the function has rational zeros, then the zeros will be factors of the constant term.

The factors of the constant term 8 are ± 1, ± 2, ± 4, and ± 8, so these are the only possible rational zeros for $P(x)$.

Use synthetic division to identify the zeros.

$\dfrac{p}{q}$	1	-3	-6	8	
-1	1	-4	-2	10	
1	1	-2	-8	0	1 is a zero.
-2	1	-5	4	0	-2 is a zero.
2	1	-1	-8	-8	
-4	1	-7	22	-80	
4	1	1	-2	0	4 is a zero.

The zeros of the function are -2, 1, and 4.

Practice Exercises

$P(x) = 3x^3 + 4x^2 - 17x - 6$

6.27   Identify the possible rational zeros of $P(x)$.

6.28   Find the rational zeros of $P(x)$.

In the previous example, where $P(x) = x^3 - 3x^2 - 6x + 8$, the zeros of the function could have been found using less synthetic division. You found that 1 is a zero and $x - 1$ is a factor. From the synthetic division table you can see that $P(x) = (x - 1)(x^2 - 2x - 8)$. You can find the remaining roots by solving the related quadratic equation $x^2 - 2x - 8 = 0$, called the *depressed equation*. You could have factored $x^2 - 2x - 8 = 0$ to make $(x - 4)(x + 2) = 0$. From the factored equation, you can identify the remaining zeros as 4 and -2.

Practice Exercises

6.29   Find the rational zeros of $P(x) = x^3 - 28x + 48$ by identifying one zero and then solving the depressed equation.

The Fundamental Theorem of Algebra

When you solved quadratic equations in Chapter 5, you found that there may be one or two real solutions or no real solutions. Recall that by completing the square or using the quadratic equation, you were able to identify complex solutions. Real numbers are a subset of complex numbers since a real number a can be expressed as a complex number $a + 0i$.

The Fundamental Theorem of Algebra states that every polynomial equation of degree 1 or more has at least one solution in the complex numbers.

Fundamental Theorem of Algebra

Every polynomial equation of degree 1 or more has at least one solution in the complex numbers.

A corollary to the Fundamental Theorem of Algebra is that a polynomial of degree n will have at most n solutions.

Corollary to the Fundamental Theorem of Algebra

Every polynomial equation of degree n has at most n solutions.

Examples:

$4x^2 - 3x + 1$ has at most two solutions

$2x^3 + x - 2$ has at most three solutions

$-3x^4 + x^3 + 5x^2 + 3x - 2$ has at most four solutions.

Note that some polynomial equations will have fewer solutions than the degree of the equation due to repeated solutions.

For example, the equation $x^2 + 2x + 1 = 0$ has only one solution since $x^2 + 2x + 1 = (x+1)^2 = (x+1)(x+1) = 0$ and -1 is a repeated solution.

Example: Solve $2x^3 + 3x^2 + 3x + 1 = 0$.

Solution: The degree of the polynomial is 3, so there are at most three solutions. Recall that solutions to the equation will also be zeros of $P(x) = 2x^3 + 3x^2 + 3x + 1$. Identify the possible rational solutions and check with synthetic division. Once you have located one rational solution, find the solution to the depressed equation using factoring, completing the square, or using the quadratic equation.

p: factors of 1: ± 1

q: factors of 2: ± 1, ± 2

$\dfrac{p}{q} = \pm 1, \ \pm \dfrac{1}{2}$

$\dfrac{p}{q}$	2	3	3	1	
-1	2	1	2	-1	
1	2	5	8	9	
$-\dfrac{1}{2}$	2	2	2	0	$-\dfrac{1}{2}$ is a solution.

The synthetic division shows that $-\dfrac{1}{2}$ is a solution to $2x^3 + 3x^2 + 3x + 1 = 0$.

Furthermore, $x - \dfrac{1}{2}$ is a factor of $2x^3 + 3x^2 + 3x + 1$ and the remaining factor is the quadratic $2x^2 + 2x + 2$. Solve the depressed equation $2x^2 + 2x + 2 = 0$ to find the remaining solutions.

$2x^2 + 2x + 2 = 0$	Depressed equation.
$x^2 + x + 1 = 0$	Divide out common factor of 2.
$x = \dfrac{-1 \pm \sqrt{1^2 - 4(1)(1)}}{2(1)}$	Use the quadratic equation.

$$= \frac{-1 \pm \sqrt{-3}}{2} \qquad \text{Simplify.}$$

$$= \frac{-1 \pm i\sqrt{3}}{2} \qquad \text{Simplify.}$$

The solutions are $-\frac{1}{2}, = \frac{-1 \pm i\sqrt{3}}{2}$.

You can check the solutions by substituting them into the original equation.

Practice Exercises

6.30   Solve. $x^3 - 3x^2 - 5x + 4 = 0$

Descartes' Rule of Signs

The Rational Zero Theorem can assist you in locating rational zeros. What about irrational zeros? Descartes' Rule of Signs can help you identify the number and kind of real zeros.

Descartes' Rule of Signs

The number of positive real zeros of a polynomial function $P(x)$ is equal to the number of sign changes of the coefficients of the function, or is less than this number by an even number. The number of negative real zeros of $P(x)$ is equal to the number of sign changes of the coefficients of $P(-x)$ or is less than this number by an even number.

Example: Determine the possible number of positive and negative real zeros of $P(x) = x^4 + 4x^3 - x^2 + 3x - 2$.

Solution: Use Descartes' Rule of Signs to evaluate the number of sign changes in the coefficients of $P(x)$ and $P(-x)$.

$$P(x) = x^4 + 4x^3 - x^2 + 3x - 2$$
$$\quad\ +\quad\ +\quad\ -\quad\ +\quad\ -$$
$$\qquad\ \ \underbrace{\quad}\underbrace{\quad}\underbrace{\quad}$$
$$\qquad\quad 1\quad 2\quad 3 \qquad \text{3 sign changes}$$

By Descartes' Rule, $P(x)$ can have three or one positive real zeros.

$$P(-x) = (-x)^4 + 4(-x)^3 - (-x)^2 + 3(-x) - 2$$

$$P(-x) = x^4 - 4x^3 - x^2 - 3x - 2$$

$$\underbrace{+ \quad - \quad - \quad - \quad -}_{1} \qquad \text{1 sign change}$$

By Descartes' Rule, $P(x)$ has one negative real zero.

$P(x)$ has one negative real zero and one or three positive real zeros.

Establishing Upper and Lower Bounds

Descartes' Rule also shows that a polynomial whose terms are all positive or all negative cannot have any positive real zeros. This fact can help you limit the area in which you look for the positive and negative real zeros that you have shown exist.

Upper and Lower Bound Theorem

For a positive number u, if $P(u)$ is divided by $(x - u)$ and the remainder and the coefficients of the quotient are all positive or all negative, then the zeros of $P(x)$ cannot be greater than u. The value of u is an upper bound of $P(x)$. If v is an upper bound for $P(-x)$, then $-v$ is a lower bound for $P(x)$.

Example: Locate upper and lower bounds for $P(x) = x^4 + 4x^3 - x^2 + 3x - 2$.

Solution: Use synthetic division to test positive integers for x to find the upper bound.

x	1	4	−1	3	−2	
1	1	5	4	7	5	Remainder and coefficients all positive.

The remainder and all of the coefficients in the quotient are positive, so 1 is an upper bound of $P(x)$.

$$P(-x) = (-x)^4 + 4(-x)^3 - (-x)^2 + 3(-x) - 2$$

$$P(-x) = x^4 - 4x^3 - x^2 - 3x - 2$$

Use synthetic division to test positive integers for x to find the lower bound.

x	1	−4	−1	−3	−2	
1	1	−3	−4	−7	−9	
2	1	−2	−5	−13	−28	
3	1	−1	−4	−15	−47	
4	1	0	−1	−7	−30	
5	1	1	4	17	83	Remainder and coefficients all positive.

The remainder and all of the coefficients in the quotient are positive, so −5 is a lower bound of $P(x)$.

A lower bound of $P(x)$ is −5 and an upper bound is 2.

The lower and upper bounds of a function define the interval where you need to look for real zeros. For the function in the previous example, the real zeros of $P(x)$ are between −5 and 2.

For real zeros that are not rational, you will be able to identify between which successive integers the zero lies by using the Location Theorem presented in Lesson 6.5. Recall that if $y = f(x)$ is a polynomial function and $f(a) < 0$ and $f(b) > 0$ for real numbers a and b, then there is at least one real zero between a and b.

Example: Locate between which successive integers the zeros lie for

$$P(x) = x^4 + 4x^3 - x^2 + 3x - 2.$$

Solution: The previous examples show that $P(x)$ has one negative real zero and one or three positive real zeros and that those zeros will occur between −5 and 2. Use synthetic division to test values between −5 and 0 to locate one negative real zero and between 0 and 2 to locate one or three positive real zeros. A change of sign in the remainder between two successive integers indicates that at least one zero lies between the two integers.

x	1	4	−1	3	−2	
−5	1	−1	4	−17	83	
−4	1	0	−1	7	−30	$f(x)$ changes sign between $x = -5$ and −4.

There is only one negative real zero, and it lies between −5 and −4. You do not need to check any other negative values.

x	1	4	−1	3	−2	
0	1	4	−1	3	−2	
1	1	5	4	7	5	$f(x)$ changes sign between $x = 0$ and 1.
2	1	6	11	25	48	

There is a zero that lies between −5 and −4 and one or three zeros that lie between 0 and 1. If there is only one zero between 0 and 1, then the remaining two zeros are imaginary.

The graph of the function is below. Notice that there are two real zeros where the function crosses the x-axis. The two remaining zeros are imaginary.

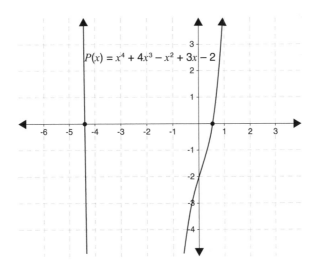

Practice Exercises

$P(x) = x^3 - 3x^2 - 5x + 4$

6.31 Identify the possible number of positive and negative real zeros of $P(x)$.

6.32 Find the upper and lower bounds of the zeros of $P(x)$.

6.33 Locate between which successive integers the zeros of $P(x)$ lie.

You have many strategies for finding zeros of a polynomial function. The following steps can help you locate the zeros of a polynomial function:

1. Determine the maximum number of zeros.

2. Use Descartes' Rule to determine the possible number of positive and negative real zeros.

3. Evaluate the upper and lower bounds of the real zeros.

4. Locate a zero using one or both of the following techniques:

   • Evaluate integer values of x within the upper and lower bounds to find a zero or to find successive integers between which a zero lies.

   • Evaluate rational zero possibilities given by values of $\dfrac{p}{q}$ where p is a factor of the constant and q is a factor of the leading coefficient.

5. Write the depressed equation and locate a zero from the depressed equation using the previous techniques.

6. Continue until the depressed equation is a quadratic equation. Use quadratic techniques to solve the depressed equation.

Example: Find all the zeros of $P(x) = 2x^4 + 3x^3 - 21x^2 + 9x + 10$.

   Solution:

   Step 1: Determine the number of zeros. The degree of $P(x)$ is 4, so there are at most four zeros.

   Step 2: Use Descartes' Rule to determine the possible number of positive and negative real zeros.

$$P(x) = 2x^4 + 3x^3 - 21x^2 + 9x + 10$$

2 sign changes

There are two or no positive real zeros.

$$P(-x) = 2(-x)^4 + 3(-x)^3 - 21(-x)^2 + 9(-x) + 10$$

$$P(-x) = 2x^4 - 3x^3 - 21x^2 - 9x + 10$$

$$\begin{array}{ccccc} + & - & - & - & + \end{array}$$

$$\underbrace{\qquad}_{1} \qquad \underbrace{\qquad}_{2} \qquad \text{2 sign changes}$$

There are two or no negative real zeros.

Step 3: Evaluate the upper and lower bounds.

$$P(x) = 2x^4 + 3x^3 - 21x^2 + 9x + 10$$

x	2	3	−21	9	10	
1	2	5	−16	−7	3	
2	2	7	−7	−5	0	2 is a zero of $P(x)$.
3	2	9	6	27	91	Remainder and coefficients all positive.

An upper bound is 3.

$$P(-x) = 2x^4 - 3x^3 - 21x^2 - 9x + 10$$

x	2	−3	−21	−9	10	
1	2	−1	−22	−31	−21	
2	2	1	−19	−47	−84	
3	2	3	−12	−45	−125	
4	2	5	−1	−13	−42	
5	2	7	14	61	315	Remainder and coefficients all positive.

A lower bound is −5.

Step 4: Locate a zero. When testing for the upper bound, 2 was identified as a zero.

Step 5: Write the depressed equation and locate a zero.

From the synthetic division table, when $x = 2$, the depressed equation is $2x^3 + 7x^2 - 7x - 5 = 0$.

$p = \pm 1, \pm 5$

$q = \pm 1, \pm 2$

$\dfrac{p}{q} = \pm 1, \pm \dfrac{1}{2}, \pm 5, \pm \dfrac{5}{2}$

1 was already tested when establishing the upper bound. 1 is not a zero. The zeros will occur between the values of −5 and 3.

Remaining values of $\dfrac{p}{q} = -1,\ \pm\dfrac{1}{2},\ -5,\ \pm\dfrac{5}{2}$.

$2x^3 + 7x^2 - 7x - 5 = 0$

$\dfrac{p}{q}$	2	7	−7	−5
−1	2	5	−12	7
$-\dfrac{1}{2}$	2	6	−10	0

$-\dfrac{1}{2}$ is a zero of $P(x)$.

Step 6: Continue until the depressed equation is a quadratic equation. Solve using quadratic equation techniques.

From the last line of the synthetic division, the depressed equation is $2x^2 + 6x - 10 = 0$. Dividing both sides of the equation by 2 yields $x^2 + 3x - 5 = 0$.

Using the quadratic equation gives:

$$x = \frac{-3 \pm \sqrt{(3)^2 - 4(1)(-5)}}{2(1)}$$

$$= \frac{-3 \pm \sqrt{29}}{2}$$

$\dfrac{-3 \pm \sqrt{29}}{2}$ are zeros of $P(x)$.

The zeros of $P(x)$ are $2,\ -\dfrac{1}{2},\ \dfrac{-3+\sqrt{29}}{2}$, and $\dfrac{-3-\sqrt{29}}{2}$.

Practice Exercises

Find all zeros of $P(x)$.

6.34 $P(x) = x^4 + 3x^3 + 4x^2 - 8$

Chapter 7

Radical Functions and Relations

7.1 Square Root Function

If a function includes the square root of a variable, it is called a *square root function*. The parent function is $f(x) = \sqrt{x}$. The domain of the parent function is $x \geq 0$ since the square root of a negative number is not defined over the set of real numbers. Furthermore, the range of the parent function is $f(x) \geq 0$ since the square root of a number is positive.

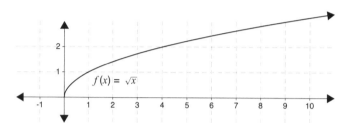

$f(x) = \sqrt{x}$

Domain: $x \geq 0$

Range: $f(x) \geq 0$

Notice that as x approaches the minimum value of 0, so does $f(x)$. As x gets larger, so does $f(x)$.

The Domain and Range of Square Root Functions

For square root functions other than the parent function, you must determine the values for which the function is defined.

Example: State the domain and the range of $f(x) = \sqrt{x - 3}$.

> **Solution:** The domain includes values for which the expression under the radical is greater than or equal to 0.
>
> $x - 3 \geq 0$
>
> $x \geq 3$
>
> The domain is $x \geq 3$.
>
> The range includes all possible values after the radical is applied to the expression. The minimum value of $f(x)$ occurs at the minimum value of x.
>
> $f(3) = \sqrt{3 - 3} = \sqrt{0} = 0$
>
> The range is $f(x) \geq 0$.

Example: State the domain and the range of $f(x) = \sqrt{x} + 4$.

> **Solution:** The domain includes values for which the expression under the radical is greater than or equal to 0.
>
> $x \geq 0$
>
> The domain is $x \geq 0$.
>
> The range includes all possible values after the radical is applied to the expression. The minimum value of $f(x)$ occurs at the minimum value of x.
>
> $f(0) = \sqrt{0} + 4 = 4$
>
> The range is $f(x) \geq 4$.

Practice Exercises

State the domain and range of each function.

7.1 $f(x) = 2\sqrt{x+1}$

7.2 $f(x) = \sqrt{x-2} - 3$

Graphing Square Root Functions Using Transformations

In Lesson 4.4 you saw how to use transformations to graph the quadratic and absolute value functions. These same transformations can be applied to square root functions. Recall that you should apply transformations to the parent function in the same order that you would evaluate the function using the order of operations.

Square Root Function: $f(x) = a\sqrt{x-h} + k$

h translates the square root function left or right.

- If $h > 0$, the function is translated h units to the right.
- If $h < 0$, the function is translated $|h|$ units to the left.

a reflects and dilates the square root function.

- If $a < 0$, the function is reflected over the x-axis.
- If $|a| > 1$, the function is stretched vertically.
- If $0 < |a| < 1$, the function is compressed vertically.

k translates the square root function up or down.

- If $k > 0$, the function is translated k units up.
- If $k < 0$, the function is translated $|k|$ units down.

a, h, and k also determine the domain and range of the square root function.

- The domain of the function is $x \geq h$.
- The range of the function is $f(x) \geq k$ if $a > 0$ and $f(x) \leq -k$ if $a < 0$.

To graph a square root function, it can be helpful to make a table of several values to get the correct shape of the graph.

Example: Graph $f(x) = \sqrt{x-2} + 1$.

Solution: Identify the values of h, a, and k in $f(x) = a\sqrt{x-h} + k$ and note their effect on the parent function. Make a table of several values. If desired, graph the parent function. Graph the given function using the transformations and the table of values.

$h = 2$; the function is translated 2 units to the right.

$a = 1$; the function is not reflected or dilated.

$k = 1$; the function is translated 1 unit up.

Domain: $x \geq 2$

Range: $f(x) \geq 1$

x	$\sqrt{x-2}+1$	$f(x)$
2	$\sqrt{2-2}+1$	1
3	$\sqrt{3-2}+1$	2
6	$\sqrt{6-2}+1$	3

Example: Graph $f(x) = -2\sqrt{x+4} - 1$.

Solution: Identify the values of h, a, and k in $f(x) = a\sqrt{x-h} + k$ and note their effect on the parent function. Make a table of several values. If desired, graph the parent function. Graph the given function using the transformations and the table of values.

$h = -4$; the function is translated 4 units to the left.

$a = -2$; the function is reflected over the x-axis and stretched vertically.

$k = -1$; the function is translated 1 unit down.

Domain: $x \geq -4$

Range: $f(x) \leq -1$

x	$-2\sqrt{x+4} - 1$	$f(x)$
-4	$-2\sqrt{-4+4} - 1$	-1
-3	$-2\sqrt{-3+4} - 1$	-3
0	$-2\sqrt{0+4} - 1$	-5

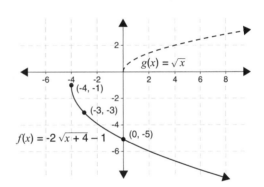

Practice Exercises

Graph each function.

7.3 $f(x) = 2\sqrt{x+1}$

7.4 $f(x) = \sqrt{x-2} - 3$

7.2 *n*th Roots

A *radical expression* contains an expression of the form $\sqrt[n]{a}$ and is read as "the *n*th root of *a*." The expression under the radical sign is called the *radicand*, and *n*, an integer greater than 1, is the *index*. If no index is given, then it is assumed to be 2, and the radical expression is a square root.

The nth Root of a

If $x = \sqrt[n]{a}$, then $x^n = a$

Notice, then, that if $x = \sqrt{25}$, then $x^2 = 25$, and $x = 5$ or -5. There can be a positive and negative root whenever n is even. The positive root is called the *principal root* Additional notation is used to indicate the negative root or both roots.

$\sqrt{25} = 5$ The principal root will have no additional notation.

$-\sqrt{25} = -5$ The negative root is indicated by a negative sign before the radical.

$\pm\sqrt{25} = \pm 5$ Both roots are indicated by a $\pm$ sign before the radical.

Example: Simplify each of the following: $\sqrt{9}$, $\sqrt{-9}$, $\sqrt[3]{8}$, $\sqrt[3]{-8}$, $\sqrt[4]{\dfrac{1}{16}}$

Solution: Use the definition of the nth root of a.

If $x = \sqrt{9}$, then $x^2 = 9$, and $x = 3$ or -3.

The principal root is 3, so $\sqrt{9} = \sqrt{3^2} = 3$.

If $x = \sqrt{-9}$, then $x^2 = -9$. There are no real values for x.

$\sqrt{-9}$ is not a real number. Recall that in the complex number system, $\sqrt{-9} = 3i$.

If $x = \sqrt[3]{8}$, then $x^3 = 8 = 2^3$, and $x = 2$.

So, $\sqrt[3]{8} = \sqrt[3]{2^3} = 2$.

If $x = \sqrt[3]{-8}$, then $x^3 = -8 = (-2)^3$, and $x = -2$.

So, $\sqrt[3]{-8} = \sqrt[3]{(-2)^3} = -2$.

If $x = \sqrt[4]{\dfrac{1}{16}}$, then $x^4 = \dfrac{1}{16} = \left(\dfrac{1}{2}\right)^4$, and $x = \dfrac{1}{2}$.

So, $\sqrt[4]{\dfrac{1}{16}} = \sqrt[4]{\left(\dfrac{1}{2}\right)^4} = \dfrac{1}{2}$.

When a radicand is raised to the same power as the root, the exponent and the root "cancel" each other out.

$$\sqrt{3^2} = 3, \ \sqrt[3]{2^3} = 2, \ \sqrt[3]{(-2)^3} = -2, \ \sqrt[4]{\left(\dfrac{1}{2}\right)^4} = \dfrac{1}{2}$$

Simplifying radical expressions will follow the general principles outlined in the following table.

n	a	$\sqrt[n]{a}$	Examples
even	positive	positive	$\sqrt[4]{81} = \sqrt[4]{3^4} = 3$ principal root
	negative	not real; complex	$\sqrt[4]{-81}$ not real; $\sqrt[4]{-81} = 3i$
	zero	zero	$\sqrt[4]{0} = 0$
odd	positive	positive	$\sqrt[3]{125} = \sqrt[3]{5^3} = 5$
	negative	negative	$\sqrt[3]{-125} = \sqrt[3]{(-5)^3} = -5$
	zero	zero	$\sqrt[3]{0} = 0$

Practice Exercises

Simplify.

7.5 $\sqrt[3]{64}$

7.6 $\sqrt{-4}$

7.7 $\sqrt[3]{-64}$

7.8 $\sqrt[4]{256}$

You may have to use absolute value signs if you take the root of a variable.

- If the index is odd, never use absolute value signs. Recall that $x = x^1$ has an odd exponent.

- If the index is even, the absolute value signs are used if you need to ensure the solution represents the principal, or non-negative, root. This is necessary if the exponent of the radicand is even and the exponent of the root is odd.

Example: Simplify $\sqrt{9x^4y^6}$, $\sqrt[3]{-8x^3y^6}$, $\sqrt[4]{81x^8y^8z^{16}}$.

> **Solution:** Rewrite each expression with factors whose exponent is the same as the index. Simplify.
>
> $$\sqrt{9x^4y^6} = \sqrt{(3)^2(x^2)^2(y^3)^2} = 3x^2|y|^3$$
>
> y^3 will be negative if $y < 0$. Therefore, the solution would be negative and would not represent the principal root. Furthermore, the index and the exponent of y are even and the exponent of the root is odd. Use absolute value signs around y.
>
> $$\sqrt{9x^4y^6} = 3x^2|y|^3$$
>
> $$\sqrt[3]{-8x^3y^6} = \sqrt[3]{(-2)^3(x)^3(y^2)^3} = -2xy^2$$
>
> Since the index is odd, absolute value signs are not needed.
>
> $$\sqrt[3]{-8x^3y^6} = -2xy^2$$
>
> $$\sqrt[4]{81x^8y^8z^{16}} = \sqrt[4]{3^4(x^2)^4(y^2)^4(z^4)^4} = 3x^2y^2z^4$$
>
> x^2, y^2, and z^4 will be positive for all values of x, y, and z. Furthermore, although the index and the exponents of the variables are even, the exponents of the roots are also even. Absolute value signs are not needed.
>
> $$\sqrt[4]{81x^8y^8z^{16}} = 3x^2y^2z^4$$

Make sure you use the properties of exponents as you rewrite and simplify radical expressions. The properties of exponents are listed in Lesson 6.1.

Practice Exercises

Simplify.

7.9 $\sqrt{16x^2y^4z^2}$

7.10 $\sqrt[3]{27x^3y^6}$

Sometimes the radicand has factors that can be simplified even though the entire radicand cannot. The product rule for radicals allows you to separate the radical into a product and simplify.

Product Rule for Radicals

For all real numbers a and b, and $n > 1$, $\sqrt[n]{ab} = \sqrt[n]{a} \cdot \sqrt[n]{b}$ if n is even and a and $b \geq 0$ or if n is odd.

Examples: $\sqrt[3]{8x} = \sqrt[3]{8} \cdot \sqrt[3]{x} = 2\sqrt[3]{x}$ and $\sqrt{10} \cdot \sqrt{2} = \sqrt{20} = \sqrt{4} \cdot \sqrt{5} = 2\sqrt{5}$

Don't forget to use absolute value signs when needed!

Use absolute value signs if the index and the exponent of the radicand are even and the exponent of the root is odd.

$$\sqrt{x^3y^2} = \sqrt{x^2y^2}\sqrt{x} = x|y|\sqrt{x}$$

Absolute value signs are not needed for x. The exponent of x in the original expression was odd. If $x < 0$, the original problem would be undefined since $x^3 < 0$. Absolute value signs are needed for y since the exponent of y in the original expression was even and the exponent of the root is odd. If $y < 0$, the original problem is still defined since $y^2 \geq 0$, but the solution will not represent the principal square root.

Example: Simplify $\sqrt{12x^3y^4z}$, $\sqrt[3]{16x^5y^2z^6}$.

Solution: Rewrite each expression with factors whose exponent is the same as the index. Simplify. Use absolute value signs if needed.

$$\sqrt{12x^3y^4z} = \sqrt{(2^2)3x^2x(y^2)^2 z} = \sqrt{(2xy^2)^2} \cdot \sqrt{3xz} = 2xy^2\sqrt{3xz}$$

If $x < 0$, the original expression would be undefined. Furthermore, the exponent of x in the original radicand is odd, so no absolute value signs are needed for x. y^2 will be positive for all values of y. The index and the exponent of y are even, but the exponent of the root is also even. Absolute value signs are not needed for y.

$$\sqrt{12x^3y^4z} = 2xy^2\sqrt{3xz}$$

$$\sqrt[3]{16x^5y^2z^6} = \sqrt[3]{\left(2^3\right)2x^3x^2y^2\left(z^2\right)^3} = \sqrt[3]{\left(2xz^2\right)^3} \cdot \sqrt[3]{2x^2y^2} = 2xz^2\sqrt[3]{2x^2y^2}$$

Since the index is odd, absolute value signs are not needed.

$$\sqrt[3]{16x^5y^2z^6} = 2xz^2\sqrt[3]{2x^2y^2}$$

Practice Exercises

Simplify.

7.11 $\sqrt[4]{48x^5y^2z^4}$

7.12 $\sqrt[3]{-54x^4y^3z}$

7.3 Working with Radical Expressions

Operations with radical expressions are similar to operations with variables. To add and subtract radical expressions, you must have like radical expressions just as you must have like terms. To multiply radical expressions you use the Distributive Property just as you do when multiplying polynomial expressions.

Adding and Subtracting Radical Expressions

Only like radical expressions can be combined by addition or subtraction. *Like radical expressions* have the same indexes and the same radicands.

Like Radical Expressions $\sqrt[4]{2x}$, $-8\sqrt[4]{2x}$ Indexes = 4, Radicands = $2x$

Unlike Radical Expressions $\sqrt[4]{2x}$, $-8\sqrt[4]{3x}$ Indexes = 4, Radicands = $2x$, $3x$

Example: Simplify $\sqrt{2} + 5 - 3\sqrt{2} + 7\sqrt{2}$.

Solution: Identify like radical expressions and combine.

$\sqrt{2} + 5 - 3\sqrt{2} + 7\sqrt{2}$

$\left(\sqrt{2} - 3\sqrt{2} + 7\sqrt{2}\right) + 5$ Identify like radical expressions.

$5\sqrt{2} + 5$ Combine like radical expressions.

If possible, simplify each radical expression before checking for like radical expressions.

Example: Simplify $\sqrt{27} + \sqrt{12} - 5\sqrt{3}$.

Solution: Simplify each radical expression. Identify like radical expressions and simplify.

$\sqrt{27} + \sqrt{12} - 5\sqrt{3}$

$\sqrt{9 \cdot 3} + \sqrt{4 \cdot 3} - 5\sqrt{3}$ Identify perfect square factors.

$\sqrt{9} \cdot \sqrt{3} + \sqrt{4} \cdot \sqrt{3} - 5\sqrt{3}$ Use the product rule for radicals.

$3\sqrt{3} + 2\sqrt{3} - 5\sqrt{3}$ Simplify radical expressions.

0 Combine like radical expressions.

Practice Exercises

Simplify.

7.13 $6\sqrt{2} - \sqrt{8} + \sqrt{50}$

7.14 $\sqrt{20} + \sqrt{45} - 3\sqrt{5}$

Multiplying Radical Expressions

If two radicals have the same index, you can multiply the radicands. After you multiply, you may be able to simplify the radical.

$\sqrt[3]{4} \cdot \sqrt[3]{16} = \sqrt[3]{64} = \sqrt[3]{4^3} = 4$

$\sqrt{2} \cdot \sqrt{6} = \sqrt{12} = \sqrt{4 \cdot 3} = \sqrt{4} \cdot \sqrt{3} = 2\sqrt{3}$

To multiply radical expressions, use the Distributive Property just like when you multiply polynomial expressions.

Multiplying Polynomial Expressions

$$(x+3)(x+2) = x^2 + 2x + 3x + 6$$
$$= x^2 + 5x + 6$$

Multiplying Radical Expressions

$$\left(\sqrt{7}+3\right)\left(\sqrt{5}+2\right) = \sqrt{7}\cdot\sqrt{5} + 2\sqrt{7} + 3\sqrt{5} + 6$$
$$= \sqrt{35} + 2\sqrt{7} + 3\sqrt{5} + 6$$

Remember that to add or subtract radical expressions, you must have like radical expressions.

Example: Simplify $\left(\sqrt[3]{4} + 2\sqrt[3]{6}\right)\left(6 - \sqrt[3]{2}\right)$.

    Solution: Distribute the first radical expression into the second and simplify.

$$\left(\sqrt[3]{4} + 2\sqrt[3]{6}\right)\left(6 - \sqrt[3]{2}\right)$$

$6\sqrt[3]{4} - \sqrt[3]{4}\cdot\sqrt[3]{2} + 12\sqrt[3]{6} - 2\sqrt[3]{6}\cdot\sqrt[3]{2}$ Use the Distributive Property.

$6\sqrt[3]{4} - \sqrt[3]{8} + 12\sqrt[3]{6} - 2\sqrt[3]{12}$ Multiply radical expressions.

$6\sqrt[3]{4} - 2 + 12\sqrt[3]{6} - 2\sqrt[3]{12}$ Simplify.

Practice Exercises

Multiply.

7.15 $\left(\sqrt{3}+1\right)\left(\sqrt{6}-2\right)$

7.16 $\left(\sqrt{2}+\sqrt{10}\right)^2$

Dividing Radical Expressions

To divide radical expressions, use the quotient rule.

Quotient Rule for Radicals

For all real numbers a and b, and $n > 1$, $\sqrt[n]{\dfrac{a}{b}} = \dfrac{\sqrt[n]{a}}{\sqrt[n]{b}}$ if n is even and a and $b \geq 0$ or if n is odd.

Examples: $\sqrt[3]{\dfrac{x}{8}} = \dfrac{\sqrt[3]{x}}{\sqrt[3]{8}} = \dfrac{\sqrt[3]{x}}{2}$ and $\dfrac{\sqrt{10}}{\sqrt{2}} = \sqrt{\dfrac{10}{2}} = \sqrt{5}$

Example: Simplify each radical expression. $\sqrt[3]{\dfrac{5}{27}}, \dfrac{\sqrt{18}}{\sqrt{3}}$

Solution: Use the quotient rule for radicals.

$$\sqrt[3]{\dfrac{5}{27}} = \dfrac{\sqrt[3]{5}}{\sqrt[3]{27}} = \dfrac{\sqrt[3]{5}}{\sqrt[3]{3^3}} = \dfrac{\sqrt[3]{5}}{3}$$

$$\dfrac{\sqrt{18}}{\sqrt{3}} = \sqrt{\dfrac{18}{3}} = \sqrt{6}$$

Practice Exercises

Simplify.

7.17 $\sqrt[4]{\dfrac{x}{81}}$

7.18 $\dfrac{\sqrt[3]{40}}{\sqrt[3]{5}}$

Rationalizing the Denominator

It is generally not accepted practice to leave a radical in the denominator of a fraction or to have a radicand that is a fraction. The process of eliminating a radical in the denominator of a fraction is called *rationalizing the denominator*. To rationalize the denominator, multiply the numerator and denominator of the fraction by a number that will eliminate the radical in the denominator.

When the denominator has a radical expression that is a single term, rationalize the denominator by multiplying the numerator and denominator by a radical that will make a perfect root in the denominator.

Example: Simplify $\sqrt{\dfrac{5}{x}}$.

Solution: Use the quotient rule. Rationalize the denominator by multiplying the numerator and denominator by $\sqrt{x}$ so that there will be an exact root in the denominator.

$$\sqrt{\dfrac{5}{x}} = \dfrac{\sqrt{5}}{\sqrt{x}} \qquad\qquad \text{Use the quotient rule for radicals.}$$

$$= \dfrac{\sqrt{5}}{\sqrt{x}} \cdot \dfrac{\sqrt{x}}{\sqrt{x}} \qquad\qquad \text{Rationalize the denominator.}$$

$$= \dfrac{\sqrt{5x}}{\sqrt{x^2}} \qquad\qquad \text{Use the product rule for radicals.}$$

$$= \dfrac{\sqrt{5x}}{x} \qquad\qquad \text{Simplify.}$$

Example: Simplify $\sqrt[4]{\dfrac{4x^6}{5y^3}}$.

Solution:

$$\sqrt[4]{\dfrac{4x^6}{5y^3}} = \dfrac{\sqrt[4]{4x^6}}{\sqrt[4]{5y^3}} \qquad\qquad \text{Use the quotient rule for radicals.}$$

$$= \dfrac{\sqrt[4]{x^4}\sqrt[4]{4x^2}}{\sqrt[4]{5y^3}} \qquad\qquad \text{Use the product rule for radicals.}$$

$$= \dfrac{\left|x\right|\sqrt[4]{4x^2}}{\sqrt[4]{5y^3}} \qquad\qquad \text{Simplify the numerator.}$$

$$= \dfrac{\left|x\right|\sqrt[4]{4x^2}}{\sqrt[4]{5y^3}} \cdot \dfrac{\sqrt[4]{5^3 y}}{\sqrt[4]{5^3 y}} \qquad\qquad \text{Rationalize the denominator.}$$

$$= \dfrac{\left|x\right|\sqrt[4]{500x^2 y}}{\sqrt[4]{5^4 y^4}} \qquad\qquad \text{Use the product rule for radicals.}$$

$$= \dfrac{\left|x\right|\sqrt[4]{500x^2 y}}{5y} \qquad\qquad \text{Simplify.}$$

Absolute value signs are needed to guarantee x is non-negative. If $x < 0$, the original expression would be defined, but the simplified expression would be negative. Note that the index and the exponent of the radicand are even, and the exponent of the root is odd.

Absolute value signs are not needed for y because if $y < 0$, the original expression would be undefined. Furthermore, in the original radicand, the exponent of y is odd.

Practice Exercises

Simplify.

7.19 $\sqrt[3]{\dfrac{8}{x}}$

7.20 $\sqrt{\dfrac{16x^4}{7y}}$

Conjugates

If a and b are not like terms, then the expressions $a + b$ and $a - b$ are *conjugates*. Recall from your work with polynomials that $(a+b)(a-b) = a^2 - b^2$. Conjugates can be used to rationalize denominators made up of two terms with one or more square roots because the product will be a rational number.

Conjugates

$2+\sqrt{5}$ and $2-\sqrt{5}$ are conjugates.

$$\left(2+\sqrt{5}\right)\left(2-\sqrt{5}\right) = 2^2 - \left(\sqrt{5}\right)^2 = 4 - 5 = -1$$

Conjugates can only be used to rationalize a denominator with a square root.

Example: Simplify $\dfrac{3-\sqrt{2}}{4+\sqrt{2}}$.

Solution: Rationalize the denominator by multiplying the numerator and the denominator by $4-\sqrt{2}$, the conjugate of the denominator. Simplify.

$$\frac{3-\sqrt{2}}{4+\sqrt{2}} = \frac{3-\sqrt{2}}{4+\sqrt{2}} \cdot \frac{4-\sqrt{2}}{4-\sqrt{2}}$$ Multiply by the conjugate.

$$= \frac{\left(3-\sqrt{2}\right)\left(4-\sqrt{2}\right)}{4^2-\left(\sqrt{2}\right)^2}$$ Multiply.

$$= \frac{12-3\sqrt{2}-4\sqrt{2}+\left(\sqrt{2}\right)^2}{16-2}$$ Multiply radical expressions.

$$= \frac{14-7\sqrt{2}}{14} = 1-\frac{\sqrt{2}}{2}$$ Simplify.

Practice Exercises

7.21   What is the conjugate of $4\sqrt{2}+2\sqrt{5}$?

7.22   Simplify $\dfrac{\sqrt{2}+3}{\sqrt{3}-\sqrt{2}}$.

7.4 Rational Exponents

Radicals can also be expressed using exponents.

If n is a natural number greater than 1 and $\sqrt[n]{b}$ is a real number, then

$$b^{\frac{1}{n}} = \sqrt[n]{b} .$$

The properties of exponents hold true, so $b^{-\frac{1}{n}} = \dfrac{1}{\sqrt[n]{b}} .$

You know that $\left(\sqrt{5}\right)^2 = 5$. Using exponents, $\left(\sqrt{5}\right)^2 = \left(5^{\frac{1}{2}}\right)^2 = \left(5\right)^{\frac{1}{2}(2)} = 5^1 = 5$.

Example: Simplify each expression. $9^{\frac{1}{2}}$, $\left(40x^5y^2\right)^{\frac{1}{3}}$, $\left(81x^4\right)^{-\frac{1}{4}}$

Solution: Change each expression into a radical and simplify.

$$9^{\frac{1}{2}} = \sqrt{9} = 3$$

$$\left(40x^5y^2\right)^{\frac{1}{3}} = \sqrt[3]{40x^5y^2} = \sqrt[3]{2^3x^3} \cdot \sqrt[3]{5x^2y^2} = 2x\sqrt[3]{5x^2y^2}$$

$$\left(81x^4\right)^{-\frac{1}{4}} = \frac{1}{\sqrt[4]{81x^4}} = \frac{1}{\sqrt[4]{3^4 x^4}} = \frac{1}{3|x|}$$

Note that absolute value signs are needed to guarantee x is non-negative. If $x < 0$, the original expression would be defined, but the simplified expression would be negative and would not be the principal root. Note that the index and the exponent of the radicand are even, and the exponent of the root is odd.

Rational exponents are defined by the following:

If n is a natural number greater than 1, m is an integer, and $\sqrt[n]{b}$ is a real number, then

$$b^{\frac{m}{n}} = \left(\sqrt[n]{b}\right)^m = \sqrt[n]{b^m}.$$

One way to work with expressions with rational exponents is to change the expression to a radical expression and simplify.

Example: Simplify each expression: $8^{\frac{2}{3}}$, $8^{-\frac{2}{3}}$, $(-8)^{\frac{2}{3}}$

Solution: Change the expression into a radical and evaluate.

$$8^{\frac{2}{3}} = \left(\sqrt[3]{8}\right)^2 = \left(\sqrt[3]{2^3}\right)^2 = 2^2 = 4$$

$$8^{-\frac{2}{3}} = \frac{1}{8^{\frac{2}{3}}} = \frac{1}{\left(\sqrt[3]{8}\right)^2} = \frac{1}{\left(\sqrt[3]{2^3}\right)^2} = \frac{1}{2^2} = \frac{1}{4}$$

$$(-8)^{\frac{2}{3}} = \left(\sqrt[3]{(-8)}\right)^2 = \left(\sqrt[3]{(-2)^3}\right)^2 = (-2)^2 = 4$$

Practice Exercises

Simplify.

7.23 $\left(16\right)^{-\frac{3}{4}}$

7.24 $\left(-27x^4y^5\right)^{\frac{1}{3}}$

Another way of simplifying an expression with rational exponents is to keep the rational exponents and use the properties of exponents. The properties of exponents are listed in Lesson 6.1.

An exponential expression is in simplest form when:

- All exponents are positive.

- The exponents in the denominator are positive integers.

- Fractional exponents are in lowest terms.

Example: Simplify $x^{\frac{1}{4}} \cdot x^{\frac{3}{4}}$.

Solution: Apply the Product of Powers Property of Exponents and add the exponents.

$$x^{\frac{1}{4}} \cdot x^{\frac{3}{4}} = x^{\frac{1}{4}+\frac{3}{4}} = x^{\frac{4}{4}} = x^1 = x$$

Example: Simplify $x^{-\frac{6}{4}}$.

Solution: Simplify the rational exponent, rewrite the expression with a positive exponent, and rationalize the denominator so that the exponent in the denominator is a positive integer.

$x^{-\frac{6}{4}} = x^{-\frac{3}{2}}$ Simplify the rational exponent.

$= \dfrac{1}{x^{\frac{3}{2}}}$ Rewrite the expression with a positive exponent.

$= \dfrac{1}{x^{\frac{3}{2}}} \cdot \dfrac{x^{\frac{1}{2}}}{x^{\frac{1}{2}}}$ Multiply the expression by a fraction equal to 1.

$= \dfrac{x^{\frac{1}{2}}}{x^{\frac{3}{2}+\frac{1}{2}}}$ Simplify.

$= \dfrac{x^{\frac{1}{2}}}{x^2}$ Simplify.

In the previous example, notice that the expression was multiplied by $\dfrac{x^{\frac{1}{2}}}{x^{\frac{1}{2}}}$ so that the exponent in the denominator would simplify to an integer.

Practice Exercises

Simplify.

7.25 $x^{\frac{2}{3}} \cdot x^{\frac{1}{4}}$

7.26 $x^{-\frac{3}{5}}$

Rational exponents can make it easier to simplify radical expressions as well. Final answers should be given in radical form with the lowest possible index.

Example: Simplify $\dfrac{\sqrt[8]{16}}{\sqrt[3]{2}}$.

Solution: Rewrite the expression using rational exponents. Simplify the numerator. Use the properties of exponents to simplify.

$\dfrac{\sqrt[8]{16}}{\sqrt[3]{2}} = \dfrac{16^{\frac{1}{8}}}{2^{\frac{1}{3}}}$ Rewrite the expression with rational exponents.

$= \dfrac{\left(2^4\right)^{\frac{1}{8}}}{2^{\frac{1}{3}}}$ $16 = 2^4$ and will give a common base of 2.

$= \dfrac{2^{\frac{4}{8}}}{2^{\frac{1}{3}}}$ Power of a Power Property of Exponents.

$= \dfrac{2^{\frac{1}{2}}}{2^{\frac{1}{3}}}$ Simplify the rational exponent in the numerator.

$= 2^{\frac{1}{2} - \frac{1}{3}}$ Quotient of Powers Property of Exponents.

$= 2^{\frac{3}{6} - \frac{2}{6}}$ Find a common denominator.

$= 2^{\frac{1}{6}} = \sqrt[6]{2}$ Simplify and change to radical.

Notice in the previous example that using rational expressions allowed you to combine radicals with different indexes because the bases of the rational expressions were the same after they were simplified.

A radical expression is in simplest form when:

- The index n is as small as possible.

- The radicand has no factors other than 1 that are nth powers of an integer or a polynomial.

- There are no fractions in the radicand.

- There are no radicals in the denominator.

Example: Simplify $\sqrt[7]{x^4} \cdot \sqrt[3]{x^2}$.

>**Solution:** Rewrite the expression using rational exponents. Use the properties of exponents to simplify.

$$\sqrt[7]{x^4} \cdot \sqrt[3]{x^2} = x^{\frac{4}{7}} \cdot x^{\frac{2}{3}} \qquad \text{Rewrite the expression using rational exponents.}$$

$$= x^{\frac{4}{7}+\frac{2}{3}} \qquad \text{Use the Product of Powers Exponent Property.}$$

$$= x^{\frac{12}{21}+\frac{14}{21}} \qquad \text{Find a common denominator.}$$

$$= x^{\frac{26}{21}} \qquad \text{Simplify.}$$

$$= \sqrt[21]{x^{26}} \qquad \text{Rewrite the expression using a radical.}$$

$$= \sqrt[21]{x^{21}} \cdot \sqrt[21]{x^5} \qquad \text{Use the product rule for radicals.}$$

$$= x \cdot \sqrt[21]{x^5} \qquad \text{Simplify.}$$

Practice Exercises

Simplify.

7.27 $\dfrac{\sqrt{3}}{\sqrt[5]{9}}$

7.28 $\sqrt{\left(x\right)^3} \cdot \sqrt[4]{x}$

7.5 Radical Equations

If an equation has a variable in a radicand, the equation is called a *radical equation*. If an equation has a variable raised to a rational exponent, the equation is equivalent to a radical equation.

Radical Equations	Equations Equivalent to Radical Equations
$\sqrt{x+5} = 7$	$(x+5)^{\frac{1}{2}} = 7$
$\sqrt[3]{6x^2 - 1} = 4$	$(6x^2 - 1)^{\frac{1}{3}} = 4$
$\sqrt[5]{x-2} = \sqrt[5]{x^2 + 1}$	$(x-2)^{\frac{1}{5}} = (x^2 + 1)^{\frac{1}{5}}$

To solve a radical equation:
- Isolate the radical expression on one side of the equation.
- Raise both sides of the equation to the power equal to the index of the radical.
- Solve the resulting equation.
- Check your answers in the original equation to eliminate extraneous solutions.

When you raise both sides of the equation to a power, the resulting equation will sometimes have *extraneous solutions* that are not solutions to the original equation.

Example: Solve $\sqrt[4]{x-2} - 1 = 1$.

> **Solution:** Add 1 to both sides to isolate the radical. Raise both sides to the fourth power. Solve the resulting equation. Check the solution in the original equation.

$$\sqrt[4]{x-2} - 1 = 1$$

$$\sqrt[4]{x-2} = 2 \qquad \text{Add 1 to both sides of the equation.}$$

$$\left(\sqrt[4]{x-2}\right)^4 = 2^4 \qquad \text{Raise both sides to the fourth power.}$$

$$x - 2 = 16 \qquad \text{Simplify.}$$

$$x = 18 \qquad \text{Add 2 to both sides of the equation.}$$

Check:

$\sqrt[4]{18-2}-1=1$

$\sqrt[4]{16}-1=1$

$2-1=1$ ✓

$x = 18$ is the solution.

When you raise both sides of an equation to a power, the resulting equation may be a quadratic equation.

Example: Solve $\sqrt{2x+6}+1=x$.

Solution: Isolate the radical by subtracting 1 from both sides of the equation. Square both sides of the equation. Solve the resulting quadratic equation. Check the solutions in the original equation.

$\sqrt{2x+6}+1=x$

$\sqrt{2x+6}=x-1$	Subtract 1 from both sides of the equation.
$\left(\sqrt{2x+6}\right)^2=(x-1)^2$	Square both sides of the equation.
$2x+6=x^2-2x+1$	Simplify.
$0=x^2-4x-5$	Subtract $2x + 6$ from both sides of the equation.
$0=(x-5)(x+1)$	Factor the quadratic expression.
$x = 5, -1$	Use the Zero Product Property.

Check:

$x = 5$	$x = -1$
$\sqrt{2(5)+6}+1=5$	$\sqrt{2(-1)+6}+1=-1$
$\sqrt{16}+1=5$	$\sqrt{4}+1=-1$
$4 + 1 = 5$ ✓	$2 + 1 = -1$ FALSE

$x = 5$ is the solution. -1 is an extraneous solution.

If there is more than one radical in the equation, you may need to raise both sides to a power more than once. Isolate one radical and raise both sides of the equation to a power, and then isolate the next radical and raise both sides of the equation to a power.

Example: Solve $\sqrt{4x+1} = \sqrt{x+2} + 1$.

Solution: Since one radical is isolated, square both sides. Isolate the remaining radical and square both sides again. Solve the resulting quadratic equation. Check solutions in the original equation.

$$\sqrt{4x+1} = \sqrt{x+2} + 1$$

$$\left(\sqrt{4x+1}\right)^2 = \left(\sqrt{x+2} + 1\right)^2 \qquad \text{Square both sides.}$$

$$4x+1 = x+2+2\sqrt{x+2}+1 \qquad \text{Simplify.}$$

$$4x+1 = x+3+2\sqrt{x+2} \qquad \text{Simplify.}$$

$$3x-2 = 2\sqrt{x+2} \qquad \text{Subtract } x+3 \text{ from both sides to isolate the radical.}$$

$$\left(3x-2\right)^2 = \left(2\sqrt{x+2}\right)^2 \qquad \text{Square both sides.}$$

$$9x^2 - 12x + 4 = 4(x+2) \qquad \text{Simplify.}$$

$$9x^2 - 12x + 4 = 4x + 8 \qquad \text{Distribute.}$$

$$9x^2 - 16x - 4 = 0 \qquad \text{Subtract } 4x + 8 \text{ from both sides of the equation.}$$

$$x = \frac{16 \pm \sqrt{(-16)^2 - 4(9)(-4)}}{2 \cdot 9} \qquad \text{Use the quadratic formula.}$$

$$x = \frac{16 \pm \sqrt{256 + 144}}{2 \cdot 9} \qquad \text{Simplify.}$$

$$x = \frac{16 \pm 20}{18} \qquad \text{Simplify.}$$

$$x = 2, \ -\frac{2}{9} \qquad \text{Simplify.}$$

Check:

$$x = 2 \qquad\qquad x = -\frac{2}{9}$$

$$\sqrt{4(2)+1} = \sqrt{2+2} + 1 \qquad \sqrt{4\left(-\frac{2}{9}\right)+1} = \sqrt{-\frac{2}{9}+2} + 1$$

$$\sqrt{9} = \sqrt{4} + 1 \qquad\qquad \sqrt{\frac{1}{9}} = \sqrt{\frac{16}{9}} + 1$$

$$3 = 2 + 1 \checkmark \qquad\qquad \frac{1}{3} = \frac{4}{3} + 1 \quad \text{FALSE}$$

$x = 2$ is the solution.

Practice Exercises

Solve.

7.29 $-2\sqrt[3]{2x+10} = 4$

7.30 $\sqrt{x+3} = x + 3$

7.31 $\sqrt{2x+5} = \sqrt{x+2} + 1$

To solve an equation with rational exponents, raise both sides of the equation to the power equal to the inverse of the rational exponent. Remember that to invert a fraction, switch the numerator and the denominator.

Example: Solve $(x-4)^{\frac{3}{2}} = 8$.

   Solution: Raise both sides to the $\dfrac{2}{3}$ power (the inverse of $\dfrac{3}{2}$).

   $(x-4)^{\frac{3}{2}} = 8$

   $\left((x-4)^{\frac{3}{2}}\right)^{\frac{2}{3}} = 8^{\frac{2}{3}}$ 　　　Raise both sides to the inverse of the exponent.

   $x - 4 = \left(\sqrt[3]{8}\right)^2$ 　　Simplify.

   $x - 4 = \left(\sqrt[3]{2^3}\right)^2$

   $x - 4 = (2)^2$ 　　Simplify.

   $x - 4 = 4$ 　　Simplify.

   $x = 8$ 　　Add 4 to both sides of the equation.

Check:

$$\left(8-4\right)^{\frac{3}{2}} = 8$$

$$\left(4\right)^{\frac{3}{2}} = 8$$

$$\left(\sqrt{4}\right)^{3} = 8$$

$$2^{3} = 8$$

$x = 8$ is the solution.

Practice Exercises

Solve.

7.32 $(3x+2)^{\frac{2}{5}} = 4$

Chapter 8

Exponential and Logarithmic Functions

8.1 Exponential Functions

You may have heard the term "exponential growth." This expression is typically used in everyday language to describe something that grows very rapidly. Mathematically, though, an exponential function can describe rapid growth or decrease (often called decay).

An *exponential function* has the form $f(x) = b^x$ where $b > 0$ and $b \neq 1$. Notice that the base is the constant b, while the variable is an exponent.

Example: Graph the function $f(x) = 2^x$.

> **Solution:** Substitute values for x and find $f(x)$. Graph the ordered pairs to find the general shape of the graph. Connect the points with a smooth curve.

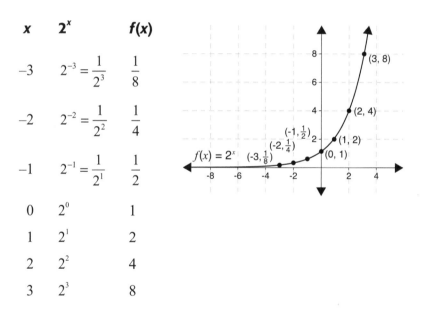

x	2^x	$f(x)$
-3	$2^{-3} = \dfrac{1}{2^3}$	$\dfrac{1}{8}$
-2	$2^{-2} = \dfrac{1}{2^2}$	$\dfrac{1}{4}$
-1	$2^{-1} = \dfrac{1}{2^1}$	$\dfrac{1}{2}$
0	2^0	1
1	2^1	2
2	2^2	4
3	2^3	8

Notice from the previous example that as x gets smaller, $f(x)$ approaches 0, although no value of x will make $f(x) = 0$. The x-axis is called an asymptote of the function. An *asymptote* is a line that a function approaches as it goes toward positive or negative infinity.

Notice also that as x gets larger, $f(x)$ increases without bound.

Example: Graph the function $f(x) = \left(\dfrac{1}{2}\right)^x$.

Solution: Substitute values for x and find $f(x)$. Graph the ordered pairs to find the general shape of the graph. Connect the points with a smooth curve.

x	$\left(\dfrac{1}{2}\right)^x$	$f(x)$
-3	$\left(\dfrac{1}{2}\right)^{-3} = 2^3$	8
-2	$\left(\dfrac{1}{2}\right)^{-2} = 2^2$	4
-1	$\left(\dfrac{1}{2}\right)^{-1} = 2^1$	2
0	$\left(\dfrac{1}{2}\right)^{0}$	1
1	$\left(\dfrac{1}{2}\right)^{1}$	$\dfrac{1}{2}$
2	$\left(\dfrac{1}{2}\right)^{2}$	$\dfrac{1}{4}$
3	$\left(\dfrac{1}{2}\right)^{3}$	$\dfrac{1}{8}$

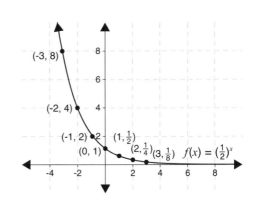

Notice in the previous example that as x gets smaller, $f(x)$ increases without bound. As x gets larger, $f(x)$ approaches 0 and the x-axis is an asymptote of the graph.

The two previous examples illustrate several properties of exponential functions.

Properties of the exponential function $f(x) = b^x$ where $b > 0$ and $b \neq 1$

- The domain is the set of all real numbers.

- The range is the set of positive real numbers.

- The y-intercept of the graph is 1.

- The x-axis is a horizontal asymptote.

- If $b > 1$, the function is increasing and is called *exponential growth*.

- If $0 < b < 1$, the function is decreasing and is called *exponential decay*.

Practice Exercises

Graph each function.

8.1 $f(x) = 3^x$

8.2 $g(x) = \left(\dfrac{2}{3}\right)^x$

The previous examples illustrate that the parent function $f(x) = b^x$ can have two different graphs depending on the value of b.

Exponential Growth   **Exponential Decay**

$f(x) = b^x$ where $b > 1$ $f(x) = b^x$ where $0 < b < 1$

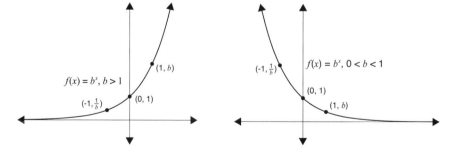

You can use transformations to graph exponential functions.

Exponential Function: $f(x) = ab^{x-h} + k$ where $b > 0$ and $b \neq 1$

h translates the function left or right.
- If $h > 0$, the function is translated h units to the right.
- If $h < 0$, the function is translated $|h|$ units to the left.

a reflects and dilates the function.
- If $a < 0$, the function is reflected over the x-axis.
- If $|a| > 1$, the function is stretched vertically.
- If $0 < |a| < 1$, the function is compressed vertically.

k translates the function up or down and identifies the asymptote.
- If $k > 0$, the function is translated k units up.
- If $k < 0$, the function is translated $|k|$ units down.
- The asymptote is the line $y = k$.

Recall that the order in which you apply the transformations mirrors the order of operations you would use to evaluate the function. First translate left or right, then dilate and reflect, and then translate up or down.

Example: Graph the function $g(x) = 3^x + 2$.

    Solution: The function is an exponential growth function since $b = 3$. In the general form $f(x) = ab^{x-h} + k$, $h = 0$, $a = 1$, and $k = 2$. The function is not translated left or right or reflected or dilated. The function is translated 2 units up from the graph of the parent function. The asymptote is the line $y = 2$. You can sketch the parent function and then translate it 2 units up.

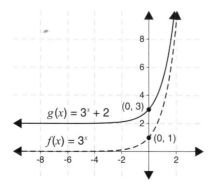

If a function has several translations from the parent function, it may be easier to substitute a few values for x to help you get the shape of the graph.

Example: Graph the function $g(x) = -\left(\dfrac{1}{2}\right)^{x-1} - 3$.

Solution: The function is an exponential decay function since $b = \dfrac{1}{2}$. In the general form $f(x) = ab^{x-h} + k$, $h = 1$, $a = -1$, and $k = -3$. Therefore the graph of the parent function is translated 1 unit to the right, reflected over the x-axis, and translated 3 units down. The asymptote is the line $y = -3$. Choose a few values for x to help you identify the shape and location of the graph.

x	$-\left(\dfrac{1}{2}\right)^{x-1} - 3$	$f(x)$
0	$-\left(\dfrac{1}{2}\right)^{0-1} - 3 = -\left(\dfrac{1}{2}\right)^{-1} - 3 = -2^1 - 3$	-5
1	$-\left(\dfrac{1}{2}\right)^{1-1} - 3 = -\left(\dfrac{1}{2}\right)^{0} - 3 = -1 - 3$	-4
2	$-\left(\dfrac{1}{2}\right)^{2-1} - 3 = -\left(\dfrac{1}{2}\right)^{1} - 3 = -\dfrac{1}{2} - 3$	$-\dfrac{7}{2}$

Practice Exercises

Graph each function.

8.3 $f(x) = 2^{x-1}$

8.4 $g(x) = 3(2)^{x+4} - 1$

8.2 Solving Exponential Equations

Growth or decay that occurs with a constant percent increase or decrease can be modeled by the exponential function

$A(t) = a(1+r)^t$ where

a is the initial value,

r is the percent increase or decrease per time period, and

t is the number of time periods.

Note that r is the percent expressed as a decimal.

Example: A college projects that their tuition and fees will rise at the rate of 8% per year. If the current tuition and fees for an incoming freshman are $10,000, what is the expected tuition and fees for that student for their fourth year of school?

Solution: Identify the values of a and r in the exponential growth function. Write the function $A(t)$. Since $A(0)$ represents the first year, evaluate $A(3)$ to find the projected tuition and fees for the fourth year.

$a = \$10,000$

$r = 8\% = 0.08$

$A(t) = \$10,000(1 + 0.08)^t$

$A(t) = \$10,000(1.08)^t$

$A(3) = \$10,000(1.08)^3 = \$12,597.12$

If the tuition and fees continue to rise at a rate of 8% per year, the projected tuition and fees for the fourth year will be about $12,600.

If the percent of change per time period represents a decrease, then the value of r is negative.

Example: A car depreciates at the rate of about 15% per year. If you purchase a car for $15,000, what is the car's value after 3 years?

> **Solution:** Identify the values of a and r in the exponential growth function. Write the function $A(t)$. Evaluate $A(3)$ to find the estimated value after 3 years.
>
> $a = \$15,000$
>
> $r = -15\% = -0.15$
>
> $A(t) = \$15,000(1 - 0.15)^t$
>
> $A(t) = \$15,000(0.85)^t$
>
> $A(3) = \$15,000(0.85)^3 = \$9,211.88$
>
> If the depreciation continues at a rate of 15% per year, the car will be worth about $9,200 after 3 years.

Practice Exercises

8.5 A business earned a profit of $5,000 in the first year. Company executives expect a 6% increase in profits each year. What will be the expected profit in each of the next 5 years?

8.6 The population of a city is decreasing at the rate of 2.5% per year. This year, the population is 105,000. What is the anticipated population in 10 years?

To solve an exponential equation with the variable as an exponent, first write both sides of the equation with the same base and apply the Equality Property of Exponential Equations.

Equality Property of Exponential Equations

> Exponential equations are one-to-one, so for $b > 0$ and $b \neq 1$,
>
> $b^m = b^n$ if and only if $m = n$.
>
> So, if $2^x = 2^4$, then $x = 4$.

Example: Solve $25 = 5^x$.

> **Solution:** Rewrite the equation so that both sides of the equation have the same base of 5. Use the Equality Property of Exponential Equations to write the equation represented by the exponents.
>
> $25 = 5^x$
>
> $5^2 = 5^x$ Use the same base b.
>
> $x = 2$ Use the Equality Property of Exponential Equations.
>
> Check: $x = 2$
>
> $25 = 5^2$
>
> $25 = 25$ ✓

Example: Solve $125 = 5^{2x-3}$.

> **Solution:** Rewrite the equation so that both sides of the equation have the same base of 5. Use the Equality Property of Exponential Equations to write the equation represented by the exponents.
>
> $125 = 5^{2x-3}$
>
> $5^3 = 5^{2x-3}$ Use the same base b.
>
> $3 = 2x - 3$ Use the Equality Property of Exponential Equations.
>
> $6 = 2x$ Add 3 to both sides of the equation.
>
> $3 = x$ Divide both sides of the equation by 2.
>
> Check: $x = 3$
>
> $125 = 5^{2(3)-3}$
>
> $125 = 5^3$
>
> $125 = 125$ ✓

Example: Solve $16^{-x+3} = 2^{2x+7}$.

> **Solution:** Rewrite the equation so that both sides of the equation have the same base of 2. Use the Equality Property of Exponential Equations to write the equation represented by the exponents.

$$16^{-x+3} = 2^{2x+7}$$

$$2^{4(-x+3)} = 2^{2x+7}$$ Use the same base b.

$4(-x + 3) = 2x + 7$ Use the Equality Property of Exponential Equations.

$-4x + 12 = 2x + 7$ Distribute.

$-6x = -5$ Subtract $2x$ from both sides of the equation.

$$x = \frac{5}{6}$$ Divide both sides of the equation by -6.

Check: $x = \dfrac{5}{6}$

$$16^{-\frac{5}{6}+3} = 2^{2\left(\frac{5}{6}\right)+7}$$

$$16^{\frac{13}{6}} = 2^{\frac{52}{6}}$$

$$2^{4\left(\frac{13}{6}\right)} = 2^{\frac{52}{6}}$$

$$2^{\frac{52}{6}} = 2^{\frac{52}{6}} \checkmark$$

Practice Exercises

Solve.

8.7 $1{,}024 = 4^x$

8.8 $27 = 3^{4x+1}$

8.9 $64^{2x-3} = 16^{x-2}$

8.3 Logarithmic Functions

Recall from Lesson 4.6 that to find the inverse of a function you switch the x and y variables. So, the inverse of the exponential function $y = b^x$ is $x = b^y$. Recall also that the graphs of inverse functions are symmetric about the line $y = x$.

Graph of $y = b^x$ and $x = b^y$
when $b > 0$

Graph of $y = b^x$ and $x = b^y$
when $0 < b < 1$

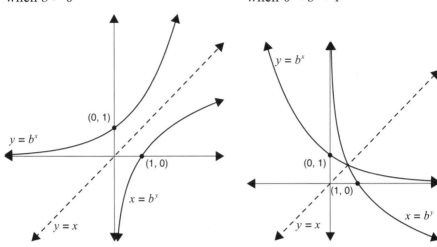

Typically, after the variables are switched to find the inverse, the equation is solved for y. In other words, we need an equation that expresses "y is the exponent of b to equal x. The logarithm function was created for this situation.

Defining the Logarithmic Function

The *logarithmic function* $y = \log_b x$ is defined as $x = b^y$, where $b > 0$ and $b \neq 1$, and is read "y equals the logarithm (log), base b, of x" or "y equals the logarithm (log) of x base b."

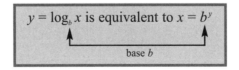

$y = \log_b x$ is equivalent to $x = b^y$

base b

Properties of the logarithmic function $f(x) = \log_b x$ where $b > 0$ and $b \neq 1$

- The domain is the set of all positive real numbers.
- The range is the set of real numbers.
- The x-intercept of the graph is 1.
- The y-axis is a vertical asymptote.

Working with Exponential and Logarithmic Forms

It is important to be able to transform equations between exponential and logarithmic forms.

Example: Rewrite each equation in logarithmic form.

$$3^2 = 9 \qquad \frac{1}{16} = \left(\frac{1}{2}\right)^4 \qquad 3^{-2} = \frac{1}{9}$$

Solution: Identify the values of x, y, and b in $x = b^y$ and substitute them into the logarithmic function $y = \log_b x$.

$3^2 = 9$; $x = 9$, $y = 2$, $b = 3$

$3^2 = 9$ is equivalent to $2 = \log_3 9$

$$\frac{1}{16} = \left(\frac{1}{2}\right)^4; \ x = \frac{1}{16}, \ y = 4, \ b = \frac{1}{2}$$

$\dfrac{1}{16} = \left(\dfrac{1}{2}\right)^4$ is equivalent to $\log_{\frac{1}{2}} \dfrac{1}{16} = 4$

$$3^{-2} = \frac{1}{9}; \ x = \frac{1}{9}, \ y = -2, \ b = 3$$

$3^{-2} = \dfrac{1}{9}$ is equivalent to $-2 = \log_3 \dfrac{1}{9}$

Example: Rewrite each equation in exponential form.

$$6 = \log_2 64 \qquad \log_4 \frac{1}{16} = -2 \qquad 3 = \log_5 125$$

Solution: Identify the values of x, y, and b in $y = \log_b x$ and substitute them into $x = b^y$.

$6 = \log_2 64$; $x = 64$, $y = 6$, $b = 2$

$6 = \log_2 64$ is equivalent to $2^6 = 64$.

$$\log_4 \frac{1}{16} = -2; \ x = \frac{1}{16}, \ y = -2, \ b = 4$$

$$\log_4 \frac{1}{16} = -2 \text{ is equivalent to } \frac{1}{16} = 4^{-2}.$$

$3 = \log_5 125; \ x = 125, \ y = 3, \ b = 5$

$3 = \log_5 125$ is equivalent to $5^3 = 125$.

One way to evaluate a logarithm is to write the logarithm in exponential form and solve the exponential equation.

Example: Evaluate each logarithm.

$$\log_9 81 \qquad\qquad \log_5 \frac{1}{25} \qquad\qquad \log_3 \sqrt{3}$$

Solution: Rewrite each logarithm in exponential form.

Let $x = \log_9 81$ Let $x = \log_5 \dfrac{1}{25}$ Let $x = \log_3 \sqrt{3}$

$$9^x = 81 \qquad\qquad 5^x = \frac{1}{25} \qquad\qquad 3^x = \sqrt{3}$$

$$9^x = 9^2 \qquad\qquad 5^x = \left(\frac{1}{5}\right)^2 = 5^{-2} \qquad\qquad 3^x = 3^{\frac{1}{2}}$$

$$x = 2 \qquad\qquad x = -2 \qquad\qquad x = \frac{1}{2}$$

$$\log_9 81 = 2 \qquad\qquad \log_5 \frac{1}{25} = -2 \qquad\qquad \log_3 \sqrt{3} = \frac{1}{2}$$

Practice Exercises

8.10   Rewrite $4^3 = 64$ as a logarithm.

8.11   Rewrite $5 = \log_2 32$ in exponential form.

8.12   Evaluate $\log_6 36$.

Solving Simple Logarithmic Equations

Logarithmic equations are equations that contain one or more logarithms. You can rewrite a simple logarithmic equation in exponential form to solve.

Remember that

$y = \log_b x$ is defined as $x = b^y$, where $x > 0$, $b > 0$ and $b \neq 1$

Example: Solve $\log_x 64 = 6$.

Solution: Rewrite the logarithm in exponential form and solve.

$\log_x 64 = 6$

$x^6 = 64$ Rewrite in exponential form.

$x^6 = 2^6$ Rewrite 64 using a base raised to the same exponent as x.

$\left(x^6\right)^{\frac{1}{6}} = \left(2^6\right)^{\frac{1}{6}}$ Raise both sides of the equation to the $\frac{1}{6}$ power.

$x = 2$ Simplify.

Check: $x = 2$

$\log_x 64 = 6$

$2^6 = 64$

$64 = 64$ ✓

Example: Solve $\log_4 x = 5$.

Solution: Rewrite the logarithm in exponential form and solve.

$\log_4 x = 5$

$4^5 = x$ Rewrite in exponential form.

$1{,}024 = x$ Simplify.

Check: $x = 1{,}024$

$\log_4 1{,}024 = 5$

$4^5 = 1{,}024$

$1{,}024 = 1{,}024$ ✓

Example: Solve $\log_3 81 = 2x - 4$.

    Solution: Rewrite the logarithm in exponential form and solve.

$\log_3 81 = 2x - 4$

    $3^{2x-4} = 81$ Rewrite in exponential form.

    $3^{2x-4} = 3^4$ Use the same base b.

    $2x - 4 = 4$ Use the Equality Property of Exponential Equations.

        $2x = 8$ Add 4 to both sides of the equation.

          $x = 4$ Divide both sides of the equation by 2.

Check: $x = 4$

$\log_3 81 = 2(4) - 4$

$\log_3 81 = 4$

    $3^4 = 81$ ✓

Practice Exercises

Solve.

8.13 $\log_x 125 = 3$

8.14 $\log_6 x = 3$

8.15 $\log_5 625 = 2x + 1$

8.4 Properties of Logarithms

Logarithms are exponents, so you will see similarities among logarithmic and exponential properties.

Properties of Logarithms

For any positive real numbers *m*, *n*, and *b*, where *b* ≠ 1, and for any real number *p*:

Identity Properties $\log_b b = 1$ because $b^1 = b$.

                      $\log_b 1 = 0$ because $b^0 = 1$.

Inverse Properties $\log_b b^p = p$ because $b^p = b^p$.

$b^{\log_b p} = p$ because if $y = \log_b p$, then $b^y = p$.

Product Property $\log_b mn = \log_b m + \log_b n$

Quotient Property $\log_b \dfrac{m}{n} = \log_b m - \log_b n$

Power Property $\log_b m^p = p \log_b m$

Property of Equality $\log_b m = \log_b n$ if and only if $m = n$

Applying the Properties of Logarithms

You can use the Product, Quotient, and Power properties to expand or simplify logarithmic expressions.

Example: Write each logarithm in expanded form.

$$\log_3 3x^2 \qquad \log_4 \frac{1}{x^3} \qquad \log_7 \sqrt{\frac{x^2}{y^4}}$$

Solution: Use the properties of logarithms to rewrite each expression in expanded form.

$$\log_3 3x^2 = \log_3 3 + \log_3 x^2 \qquad \text{Product Property.}$$

$$= 1 + \log_3 x^2 \qquad \text{Identity Property.}$$

$$= 1 + 2\log_3 x \qquad \text{Power Property.}$$

$$\log_4 \frac{1}{x^3} = \log_4 1 - \log_4 x^3 \qquad \text{Quotient Property.}$$

$$= 0 - \log_4 x^3 \qquad \text{Identity Property.}$$

$$= -3\log_4 x \qquad \text{Power Property.}$$

$$\log_7 \sqrt{\frac{x^2}{y^4}} = \log_7 \left(\frac{x^2}{y^4}\right)^{\frac{1}{2}} \qquad \text{Rewrite radical as exponent.}$$

$$= \log_7 \frac{x}{y^2} \qquad \text{Simplify.}$$

$$= \log_7 x - \log_7 y^2 \qquad \text{Quotient Property.}$$

$$= \log_7 x - 2\log_7 y \qquad \text{Power Property.}$$

You can use the properties of logarithms to help evaluate a logarithm.

Example: If $\log_3 4 \approx 1.3$ evaluate each logarithm.

$$\log_3 16 \qquad\qquad \log_3 12 \qquad\qquad \log_3 \frac{1}{4}$$

Solution: Use the properties of logarithms to rewrite each logarithm, focusing on factors of the known logarithm.

$$\log_3 16 = \log_3 4^2 \qquad \text{Rewrite 16 using an exponent and base 4.}$$

$$= 2\log_3 4 \qquad \text{Power Property.}$$

$$\approx 2(1.3) = 2.6 \qquad \text{Substitute and simplify.}$$

$$\log_3 12 = \log_3 3 \cdot 4 \qquad \text{Rewrite 12 using 4 as a factor.}$$

$$= \log_3 3 + \log_3 4 \qquad \text{Product Property.}$$

$$= 1 + \log_3 4 \qquad \text{Identity Property.}$$

$$\approx 1 + 1.3 = 2.3 \qquad \text{Substitute and simplify.}$$

$$\log_3 \frac{1}{4} = \log_3 1 - \log_3 4 \qquad \text{Quotient Property.}$$

$$= 0 - \log_3 4 \qquad \text{Identity Property.}$$

$$= 0 - 1.3 = -1.3 \qquad \text{Substitute and simplify.}$$

Practice Exercises

8.16   Write $\log_2 \frac{x^2}{4}$ in expanded form.

8.17   If $\log_2 6 \approx 2.585$, evaluate $\log_2 72$.

Using Logarithmic Properties to Solve Equations

You can use the Equality Property of Logarithmic Equations to solve logarithmic equations.

Equality Property of Logarithmic Equations

Logarithmic equations are one-to-one, so for $b > 0$ and $b \neq 1$,

$\log_b m = \log_b n$ if and only if $m = n$.

So, if $\log_2 x = \log_2 4$, then $x = 4$.

If you can simplify an equation so that there is a single logarithm with the same base on each side of the equation, you can use the Equality Property of Logarithmic Equations to write a new equation. Be sure to check your solution in the original equation to eliminate any extraneous solutions.

Example: Solve $\log_3 x = 6\log_3 2 - \log_3 4$.

Solution: Use the properties of logarithms to transform the right side of the equation into a single logarithm. Use the Equality Property of Logarithmic Equations to write a new equation and solve.

$\log_3 x = 6\log_3 2 - \log_3 4$

$\log_3 x = \log_3 2^6 - \log_3 4$ Power Property.

$\log_3 x = \log_3 \dfrac{64}{4}$ Quotient Property.

$\log_3 x = \log_3 16$ Simplify.

$x = 16$ Use the Equality Property of Logarithmic Equations.

Check: $x = 16$

$\log_3 16 = 6\log_3 2 - \log_3 4$

$\log_3 16 = \log_3 \dfrac{2^6}{4} = \log_3 16$ ✓

You can use the properties of logarithms to simplify logarithms in an equation into the logarithm of a single quantity. After the equation is simplified to a single logarithm, you can use exponential form to solve the equation.

Example: Solve $\log_6 x + \log_6 3 = 2$.

Solution: Use the properties of logarithms to combine the logarithms into a single logarithm. Convert to exponential form and solve.

$\log_6 x + \log_6 3 = 2$

$\qquad \log_6 3x = 2 \qquad$ Product Property.

$\qquad \quad 6^2 = 3x \qquad$ Definition of logarithm.

$\qquad \quad 36 = 3x \qquad$ Simplify.

$\qquad \quad 12 = x \qquad$ Divide both sides of the equation by 3.

Check: $x = 12$

$\log_6 12 + \log_6 3 = 2$

$\qquad \log_6 12 \cdot 3 = 2 \qquad$ Product Property.

$\qquad \log_6 36 = 2 \qquad$ Simplify.

$\qquad \quad 2 = 2 \checkmark \quad 6^2 = 36$, so $\log_6 36 = 2$.

$x = 12$ is the solution.

Example: Solve $\log_2 x + \log_2 (x - 6) = 4$.

Solution: Use the properties of logarithms to simplify the equation to a single logarithm. Convert to exponential form and solve.

$\log_2 x + \log_2 (x - 6) = 4$

$\qquad \log_2 x(x - 6) = 4 \qquad$ Product Property.

$\qquad \quad 2^4 = x(x - 6) \qquad$ Definition of logarithm.

$\qquad \quad 16 = x^2 - 6x \qquad$ Simplify.

$\qquad \quad 0 = x^2 - 6x - 16 \qquad$ Subtract 16 from both sides of the equation.

$\qquad \quad 0 = (x - 8)(x + 2) \qquad$ Factor the quadratic.

$x - 8 = 0 \ \text{ or } \ x + 2 = 0 \qquad$ Use the Zero Product Property.

$\qquad x = 8 \quad \text{ or } \quad x = -2 \qquad$ Solve.

Check:

$x = 8$ $x = -2$

$\log_2 8 + \log_2 (8 - 6) = 4$ $\log_2 -2 + \log_2 (-2 - 6) = 4$ Undefined

$\log_2 8 + \log_2 2 = 4$

$\log_2 8 \cdot 2 = 4$

$\log_2 16 = 4$

$4 = 4$ ✓

Since the logarithm of a negative number is undefined, -2 is not a solution.

$x = 8$ is the solution.

In the previous example, please note the following:

- The equation $0 = x^2 - 6x - 16$ could also have been solved by completing the square or using the quadratic formula.

- The extraneous solution -2 was not eliminated because it was negative. It was eliminated because after it was substituted into the equation, the equation contained logarithms of negative numbers. Recall that by definition, $\log_b x$ is defined for $x > 0$, $b > 0$ and $b \neq 1$.

Practice Exercises

Solve.

8.18 $\log_4 x = 5 \log_4 2 - \log_4 8$

8.19 $\log_6 x + \log_6 18 = 3$

8.20 $\log_3 x + \log_3 (x + 4) = 2$

8.5 Common Logarithms

The *common logarithm* is log base 10, or $\log_{10}$. If you see log written without the base, it is understood to be $\log_{10}$. If you look on a scientific calculator, you will find a LOG key. This also is $\log_{10}$.

The values of the common logarithm follow the powers of 10.

- $\log 10 = \log_{10} 10 = 1$ since $10^1 = 10$.

- $\log 100 = \log_{10} 100 = 2$ since $10^2 = 100$.

- $\log 1{,}000 = \log_{10} 1{,}000 = 3$ since $10^3 = 1{,}000$.

The properties of logarithms presented in Lesson 8.4 apply to common logarithms.

Change of Base Formula

To evaluate logarithms with a calculator, the logarithm needs to be a common log, or base 10. If it is not already a common logarithm, you can use the change of base formula.

Change of Base Formula

- For $b > 0$ and $b \neq 1$, and real number x,

$$\log_b x = \frac{\log x}{\log b}$$

- So, $\log_2 8 = \dfrac{\log 8}{\log 2}$.

Notice that we know that $\log_2 8 = 3$ because $2^3 = 8$. Using a calculator, you can evaluate $\dfrac{\log 8}{\log 2}$. Log $8 \approx 0.90309$ and $\log 2 \approx 0.30103$. $\dfrac{\log 8}{\log 2} = 3$. Therefore, you can verify that $\log_2 8 = \dfrac{\log 8}{\log 2} = 3$.

Example: Evaluate $\log_3 18$.

   Solution: Use the change of base formula to convert the logarithm to a common log. Evaluate the common log with a calculator.

$$\log_3 18 = \frac{\log 18}{\log 3}$$

$$\approx 2.6309$$

Practice Exercises

8.21   Evaluate $\log_4 7$.

Using Common Logarithms to Solve Exponential Equations

In Lesson 8.2, you solved exponential equations when you could write both sides of the equation using the same base.

$25 = 5^x$

$5^2 = 5^x$ Use the same base b.

$2 = x$ Use the Equality Property of Exponential Equations.

You can use common logarithms to solve equations when both sides cannot be written using the same base.

Example: Solve $25 = 3^x$.

    Solution: Take the log of both sides and then solve for x.

     $25 = 3^x$

   $\log 25 = \log 3^x$ Property of Equality for Logarithmic Equations.

   $\log 25 = x \log 3$ Power Property.

   $\dfrac{\log 25}{\log 3} = x$ Divide both sides by log 3.

      $x \approx 2.9299$ Use a calculator to evaluate.

   Check: $x = \dfrac{\log 25}{\log 3} \approx 2.9299$

   $25 = 3^{2.9299} \approx 24.9987$ ✓

Practice Exercises

8.22   Solve $82 = 4^x$.

8.6 Natural Logarithms and Base e

In addition to base 10, the other most common base, particularly in science, is base e. Like the numbers π and $\sqrt{2}$, the number e is an irrational number.

It is defined to be the number that $\left(1+\dfrac{1}{n}\right)^n$ approaches as n increases without bound. The value of e to the first ten decimal places is 2.7182818285.

The function $f(x) = e^x$ is called the *natural exponential function*. The inverse of the natural exponential function, the *natural logarithm,* is $f^{-1}(x) = \log_e x$. It is commonly notated as $\ln x$.

The natural exponential function and the natural logarithm are inverses of each other, so their graphs are reflections of each other over the line $y = x$.

The function $f(x) = e^x$ is used to model continuous exponential growth, while $f(x) = e^{-x}$ is used to model continuous exponential decay.

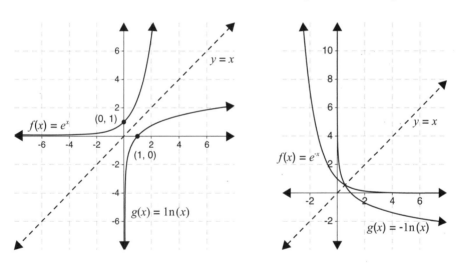

On a scientific calculator, you should find a key labeled "ln". You may have to use a second function key or inverse key to access e^x.

Example: Evaluate each of the following:

$\ln 8$ e^8

Solution: Use the ln and e^x keys on a calculator to evaluate.

$\ln 8 \approx 2.0794$

$e^8 \approx 2,980.9580$

Properties of e and ln

Since e is a real number, the properties of exponents given in Lesson 6.1 apply to expressions with e.

Properties of Exponents

For any real numbers a and b and positive integers m and n:

Product of Powers $a^m \cdot a^n = a^{m+n}$ $e^3 \cdot e^4 = e^7$

Power of a Power $\left(a^m\right)^n = a^{m \cdot n}$ $\left(e^3\right)^4 = e^{12}$

Power of a Product $\left(ab\right)^m = a^m b^m$ $\left(2e\right)^3 = 8e^3$

Power of a Quotient $\left(\dfrac{a}{b}\right)^m = \dfrac{a^m}{b^m},\ b \neq 0$ $\left(\dfrac{e}{y}\right)^2 = \dfrac{e^2}{y^2},\ y \neq 0$

Quotient of Powers $\dfrac{a^m}{a^n} = a^{m-n},\ a \neq 0$ $\dfrac{e^5}{e^2} = e^3$

Zero Exponent $a^0 = 1,\ a \neq 0$ $e^0 = 1$

Negative Exponent $a^{-m} = \dfrac{1}{a^m},\ a \neq 0$ $e^{-2} = \dfrac{1}{e^2}$

$\dfrac{1}{a^{-n}} = a^n,\ a \neq 0$ $\dfrac{1}{e^{-3}} = e^3$

Example: Simplify each expression.

$$e^4 \cdot e^2 \qquad \frac{12e^5}{3e^2} \qquad \left(2e^3\right)^4$$

Solution: Apply the rules of exponents to each expression.

$e^4 \cdot e^2 = e^{4+2} = e^6$ Product of Powers.

$\dfrac{12e^5}{3e^2} = 4e^{5-2} = 4e^3$ Quotient of Powers.

$\left(2e^3\right)^4 = 2^4 e^{3(4)} = 16e^{12}$ Power of a Power.

The properties of logarithms given in Lesson 8.4 apply to natural logarithms.

Properties of Natural Logarithms

For any postive real numbers m and $n,$ and any real number p:

Identity Properties	$\ln e = 1$	because $e^1 = e.$
	$\ln 1 = 0$	because $e^0 = 1.$
Inverse Properties	$\ln e^p = p$	because $e^p = e^p.$
	$e^{\ln p} = p$	because if $y = \ln p$, then $e^y = p.$
Product Property	$\ln mn = \ln m + \ln n$	
Quotient Property	$\ln \dfrac{m}{n} = \ln m - \ln n$	
Power Property	$\ln m^p = p \ln m$	
Property of Equality	If $\ln m = \ln n$ then $m = n$	

You can use the Product, Quotient, and Power properties to write natural logarithm expressions in expanded form.

Example: Express each logarithm in expanded form.

$$\ln 3x \qquad \ln x^4 \qquad \ln x^3 y^2$$

Solution: Use the properties of logarithms to write in expanded form.

$\ln 3x = \ln 3 + \ln x$ Product Property.

$\ln x^4 = 4 \ln x$ Power Property.

$\ln x^3 y^2 = \ln x^3 + \ln y^2$ Product Property.

$\qquad\quad = 3 \ln x + 2 \ln y$ Power Property.

You can also use the properties of logarithms to simplify natural logarithmic expressions.

Example: Write each expression as a single logarithm.

$$5 \ln x + 6 \ln y \qquad \frac{1}{2} \ln x - 2 \ln y$$

Solution: Use the properties of logarithms to simplify each expression.

$$5\ln x + 6\ln y = \ln x^5 + \ln y^6 \qquad \text{Power Property.}$$

$$= \ln x^5 y^6 \qquad \text{Product Property.}$$

$$\frac{1}{2}\ln x - 2\ln y = \ln x^{\frac{1}{2}} - \ln y^2 \qquad \text{Power Property.}$$

$$= \ln \frac{x^{\frac{1}{2}}}{y^2} = \ln \frac{\sqrt{x}}{y^2} \qquad \text{Quotient Property.}$$

Practice Exercises

8.23   Simplify $\left(\dfrac{12e^4}{6e^2}\right)^2$.

8.24   Write $\ln\sqrt{xy}$ in expanded form.

8.25   Write $2\ln 3x - 4\ln y$ as a single logarithm.

8.7 Solving Exponential and Logarithmic Equations

In Lesson 8.5, you solved exponential equations by taking the common log of both sides of the equation.

Example: Solve $25 = 3^x$.

Solution: Take the common log of both sides and then solve for x.

$$25 = 3^x$$

$\log 25 = \log 3^x$ Use the Equality Property of Logarithmic Equations.

$\log 25 = x\log 3$ Power Property.

$\dfrac{\log 25}{\log 3} = x$ Divide both sides by log 3.

$x \approx 2.9299$ Use a calculator to evaluate.

Check: $x = \dfrac{\log 25}{\log 3} \approx 2.9299$

$25 = 3^{2.9299} \approx 24.9987$ ✓

Taking the common logarithm of both sides of the equation is especially helpful if the base of the exponential equation is 10 since they are inverses.

Example: Solve $10^{2x} - 3 = 5$

Solution: Isolate the exponential and take the common log of both sides.

$10^{2x} - 3 = 5$

$\qquad 10^{2x} = 8 \qquad$ Add 3 to both sides of the equation.

$\qquad \log 10^{2x} = \log 8 \qquad$ Use the Equality Property of Logarithmic Equations.

$\qquad 2x = \log 8 \qquad$ Inverse Property.

$\qquad x = \dfrac{\log 8}{2} \qquad$ Divide both sides of the equation by 2.

$\qquad x \approx 0.4515 \quad$ Evaluate.

Check: $x = \dfrac{\log 8}{2}$

$10^{2\left(\frac{\log 8}{2}\right)} - 3 = 5$

$10^{\log 8} - 3 = 5$

$8 - 3 = 5$ ✓

You can use the inverse relationship between the natural log and the exponentials with base e to solve equations.

Example: Solve $e^x = 25$.

Solution: Take the natural logarithm of both sides of the equation.

$e^x = 25$

$\ln e^x = \ln 25 \qquad$ Use the Equality Property of Logarithmic Equations.

$\qquad x = \ln 25 \qquad$ Inverse Property.

$\qquad x \approx 3.2189 \quad$ Evaluate.

Check: $x = \ln 25$

$e^{\ln 25} = 25$

$25 = 25$ ✓

Remember to isolate e before you take the log of both sides of the equation.

Example: Solve $6e^{-3x} + 4 = 7$.

> **Solution:** Isolate e and take the natural logarithm of both sides of the equation.
>
> $6e^{-3x} + 4 = 7$
>
> $6e^{-3x} = 3$ Subtract 4 from both sides of the equation.
>
> $e^{-3x} = \dfrac{3}{6}$ Divide both sides of the equation by 6.
>
> $e^{-3x} = \dfrac{1}{2}$ Simplify.
>
> $\ln e^{-3x} = \ln \dfrac{1}{2}$ Use the Equality Property of Logarithmic Equations.
>
> $-3x = \ln \dfrac{1}{2}$ Simplify.
>
> $x = -\dfrac{1}{3} \ln \dfrac{1}{2}$ Multiply both sides of the equation by $-\dfrac{1}{3}$.
>
> $x \approx 0.2310$ Evaluate.

Check: $x = -\dfrac{1}{3} \ln \dfrac{1}{2}$

$6e^{-3\left(-\frac{1}{3} \ln \frac{1}{2}\right)} + 4 = 7$

$6e^{\ln \frac{1}{2}} + 4 = 7$

$6\left(\dfrac{1}{2}\right) + 4 = 7$

$3 + 4 = 7$ ✓

Since $\ln x$ and e^x are inverses, you can use the natural exponential function to eliminate the natural logarithm from an equation. Likewise, you can use the base 10 exponential function to eliminate a common log from an equation.

Example: Solve $2\ln 3x + 1 = 17$.

> **Solution:** Isolate the natural logarithm and use the natural exponential function to eliminate the natural logarithm.

$2\ln 3x + 1 = 17$

$2\ln 3x = 16$	Subtract 1 from both sides of the equation.
$\ln 3x = 8$	Divide both sides of the equation by 2.
$e^{\ln 3x} = e^8$	Use the Property of Equality for Exponential Functions.
$3x = e^8$	Inverse Property.
$x = \dfrac{e^8}{3}$	Divide both sides of the equation by 3.
$x \approx 993.6527$	Evaluate.

Check: $x = \dfrac{e^8}{3}$

$$2\ln 3\left(\dfrac{e^8}{3}\right) + 1 = 17$$

$$2\ln e^8 + 1 = 17$$

$$2(8) + 1 = 17$$

$$16 + 1 = 17 \checkmark$$

Example: Solve $\log 2x + 4 = 7$.

> **Solution:** Isolate the logarithm and use the Property of Equality for Exponential Functions to eliminate the logarithm.

$\log 2x + 4 = 7$

$\log 2x = 3$ Subtract 4 from both sides of the equation.

$10^{\log 2x} = 10^3$ Use the Property of Equality for Exponential Functions.

$2x = 1,000$ Inverse Property.

$x = 500$ Divide both sides of the equation by 2.

Check: $x = 500$

$\log 2(500) + 4 = 7$

$\log (1000) + 4 = 7$

$\qquad 3 + 4 = 7 \checkmark$

Practice Exercises

Solve.

8.26 $10^{2x} - 3 = 5$

8.27 $3e^{4x-2} + 1 = 7$

8.28 $6\ln 5x^2 - 2 = 4$

8.29 $4\log 2x - 3 = 1$

Chapter 9

Rational Functions and Relations

9.1 Simplify, Multiply, and Divide Rational Expressions

Recall that a rational number is a number that can be written as a ratio of two integers, or a fraction where the numerator and denominator are integers. A *rational expression* is the ratio of two polynomials, or a fraction where the numerator and denominator are polynomials. Recall that a polynomial is a general term that can refer to a monomial or an expression with many terms.

Examples of Rational Expressions

$$\frac{2}{3} \qquad \frac{2x^2y^3}{z^5} \qquad \frac{2}{x-1} \qquad \frac{x+4}{x^2-x+5} \qquad \frac{1}{x^3-2x^2+4x-6} \qquad -3x+2$$

The last example, $-3x + 2$, can be expressed as $\frac{-3x+2}{1}$, so it is still considered to be a rational expression.

One key element of working with rational expressions is remembering that the denominator of a fraction can never be equal to 0.

For example, the expression $\frac{2}{x-1}$ is not defined for $x = 1$ because if $x = 1$,

the expression would be $\frac{2}{0}$.

Simplifying Rational Expressions

A rational expression is simplified when the numerator and denominator have no common factors. To simplify a rational expression, factor the numerator and denominator completely, and then reduce common factors.

Simplifying rational expressions with variables is similar to simplifying a rational number.

Rational Number

$$\frac{12}{72} = \frac{\cancel{6} \cdot 2}{\cancel{6} \cdot 9} = \frac{2}{9}$$

Rational Expression

$$\frac{x^2-9}{x^2+6x+9} = \frac{\cancel{(x+3)}(x-3)}{\cancel{(x+3)}(x+3)} = \frac{x-3}{x+3}$$

Notice in the rational expression $\dfrac{x^2-9}{x^2+6x+9}$ that x cannot be -3 because

the denominator would be equal to 0. A number that would make the denominator of the rational expression equal to 0 is called a *restriction*.

Example: Simplify the rational expression $\dfrac{x^2-2x-15}{2x^3-50x}$ and state any restrictions on x.

> **Solution:** Factor the numerator and denominator of the fraction and reduce any common factors. Set the denominator of the original expression equal to 0 and solve to determine any restrictions on x.

$$\frac{x^2-2x-15}{2x^3-50x} = \frac{(x+3)\,\overset{1}{\cancel{(x-5)}}}{2x(x+5)\,\underset{1}{\cancel{(x-5)}}} = \frac{x+3}{2x(x+5)} \text{ or } \frac{x+3}{2x^2+10x}$$

The rational expression will be undefined if $2x^3-50x=0$.

$2x^3-50x=0$	Denominator of the original expression.
$2x(x+5)(x-5)=0$	Factor.
$2x=0,\ x+5=0$ or $x-5=0$	Use the Zero Product Property.
$x=0,\ -5,$ or 5	Solve.

So, the rational expression is undefined for $x=0,\ -5,$ or 5.

$$\frac{x^2-2x-15}{2x^3-50x} = \frac{x+3}{2x(x+5)} \text{ or } \frac{x+3}{2x^2+10x} \text{ when } x \neq 0,\ -5,\ \text{or } 5.$$

Notice that when you look for restrictions on x, you use the original rational expression, not the simplified form. If you use the simplified form of the rational expression, you will miss restrictions that are factored out.

Factoring out a -1 can help you simplify some rational expressions.

Example: Simplify the rational expression $\dfrac{2x^2 + 5x - 3}{x - 2x^2}$ and state any restrictions on x.

Solution: Factor the numerator and denominator of the rational expression and reduce any common factors. Set the denominator of the original expression equal to 0 and solve to determine any restrictions on x.

$$\frac{2x^2 + 5x - 3}{x - 2x^2} = \frac{(x+3)(2x-1)}{x(1-2x)}$$ Factor the numerator and denominator.

$$= \frac{(x+3)(-1)\,\overset{1}{\cancel{(1-2x)}}}{x\,\underset{1}{\cancel{(1-2x)}}}$$ Factor -1 out of the expression $2x - 1$.

$$= \frac{(x+3)(-1)}{x}$$ Simplify.

$$= -\frac{x+3}{x}$$ Simplify.

The rational expression will be undefined if $x - 2x^2 = 0$.

$x - 2x^2 = 0$ Denominator of original rational expression.

$x(1 - 2x) = 0$ Factor.

$x = 0$ or $1 - 2x = 0$ Use the Zero Product Property.

$x = 0$ or $\dfrac{1}{2}$ Solve.

So, the rational expression is undefined for $x = 0$ or $\dfrac{1}{2}$.

$$\frac{2x^2 + 5x - 3}{x - 2x^2} = -\frac{x+3}{x} \text{ when } x \neq 0 \text{ or } \frac{1}{2}.$$

Practice Exercises

Simplify the rational expression and state any restrictions on x.

9.1 $\dfrac{(x-1)^3}{x^2 + x - 2}$ 9.2 $\dfrac{(x-3)(x^2 + 2x - 8)}{(4 - x^2)(4 + x)}$

Multiply Rational Expressions

Multiplying rational expressions is just like multiplying rational numbers. Multiply the numerators and multiply the denominators. By leaving the numerators and denominators as the product of factors, you can easily reduce the common factors.

Rational Numbers   **Rational Expressions**

$$\frac{2}{3} \cdot \frac{3}{5} = \frac{2 \cdot \cancel{3}}{\cancel{3} \cdot 5} = \frac{2}{5} \qquad \frac{x+2}{x-3} \cdot \frac{2x(x-3)}{x+5} = \frac{(x+2) \cdot 2x \, \cancel{(x-3)}}{\cancel{(x-3)} \cdot (x+5)} = \frac{2x(x+2)}{x+5}$$

To multiply rational expressions, factor the rational expressions completely, and then multiply the numerators and the denominators, leaving them in factored form, and reduce the common factors.

Example: $\dfrac{x^2-9}{x+2} \cdot \dfrac{x^2+x-2}{3x-9}$

> **Solution:** Factor the numerator and denominator of each rational expression. Multiply the numerators and denominators and reduce the common factors.
>
> $$\frac{x^2-9}{x+2} \cdot \frac{x^2+x-2}{3x-9} = \frac{(x+3)(x-3)}{x+2} \cdot \frac{(x+2)(x-1)}{3(x-3)} \qquad \text{Factor.}$$
>
> $$= \frac{(x+3)\,\cancel{(x-3)} \cdot \cancel{(x+2)}(x-1)}{\cancel{(x+2)} \cdot 3 \, \cancel{(x-3)}} \qquad \text{Identify common factors.}$$
>
> $$= \frac{(x+3)(x-1)}{3} \quad \text{or} \quad \frac{x^2+2x-3}{3} \qquad \text{Simplify.}$$

Practice Exercises

Multiply.

9.3 $\dfrac{x^2-3x-28}{x^2-5x-14} \cdot \dfrac{6x^3+12x^2}{x^2+8x+16}$

9.4 $\dfrac{x+4}{x^2+7x+12} \cdot \dfrac{x+3}{x}$

Divide Rational Expressions

Dividing rational expressions is just like dividing rational numbers. Multiply by the reciprocal of the divisor. Recall that the divisor is the quantity that you are dividing by.

Rational Numbers   **Rational Expressions**

$$\frac{4}{7} \div \frac{4}{3} = \frac{4}{7} \cdot \frac{3}{4} = \frac{\overset{1}{\cancel{4}} \cdot 3}{7 \cdot \cancel{4}} = \frac{3}{7} \qquad \frac{x+2}{x-3} \div \frac{x+2}{x+5} = \frac{x+2}{x-3} \cdot \frac{x+5}{x+2} = \frac{\overset{1}{\cancel{(x+2)}} \cdot (x+5)}{(x-3) \cdot \cancel{(x+2)}} = \frac{x+5}{x-3}$$

> To divide rational expressions, multiply by the reciprocal of the divisor, or flip (invert) the second fraction and write the problem as a multiplication problem.

Example: $\dfrac{x^2+x-6}{x^2+4x-5} \div \dfrac{x^2-4}{x^2-1}$

Solution: Flip the divisor and rewrite as a multiplication problem. Factor the numerators and denominators. Multiply the numerators and denominators and factor out the common factors.

$$\frac{x^2+x-6}{x^2+4x-5} \div \frac{x^2-4}{x^2-1} = \frac{x^2+x-6}{x^2+4x-5} \cdot \frac{x^2-1}{x^2-4} \qquad \text{Multiply by the reciprocal.}$$

$$= \frac{(x+3)(x-2)}{(x+5)(x-1)} \cdot \frac{(x+1)(x-1)}{(x+2)(x-2)} \qquad \text{Factor.}$$

$$= \frac{(x+3)\,\cancel{(x-2)}\,(x+1)\,\cancel{(x-1)}}{(x+5)\,\cancel{(x-1)}\,(x+2)\,\cancel{(x-2)}} \qquad \text{Identify the common factors.}$$

$$= \frac{(x+3)(x+1)}{(x+5)(x+2)} \qquad \text{Simplify.}$$

Practice Exercises

Divide.

9.5　$\dfrac{3x^2-12}{4-x} \div \dfrac{x-2}{8x-32}$ 9.6　$\dfrac{x^2-2x-3}{x^2-x-6} \div \dfrac{x^2+3x+2}{x^2+4x+4}$

9.2 Add and Subtract Rational Expressions

To add and subtract rational expressions, you follow the same steps to add or subtract rational numbers. Once you have a common denominator, add or subtract only the numerators and keep the common denominator.

Rational Numbers

$$\frac{1}{5}+\frac{2}{5}=\frac{3}{5}$$

$$\frac{1}{5}+\frac{7}{10}=\frac{2}{10}+\frac{7}{10}=\frac{9}{10}$$

Rational Expressions

$$\frac{1}{x+3}+\frac{2}{x+3}=\frac{3}{x+3}$$

$$\frac{1}{x+3}+\frac{5}{(x+3)^2}=\frac{x+3}{(x+3)^2}+\frac{5}{(x+3)^2}=\frac{x+8}{(x+3)^2}$$

If the denominators are the same, add the numerators and keep the same denominator. If the denominators are not the same, you must rewrite the rational expressions with a common denominator first. After adding or subtracting the numerators, the final step is to see if the resulting rational expression can be simplified further.

Example: $\dfrac{x^2+6x-2}{x+1}-\dfrac{2x+1}{x+1}+\dfrac{6}{x+1}$

Solution: Since the denominators are the same, add and subtract the numerators as indicated by the addition and subtraction signs. Factor the resulting numerator to see if the rational expression can be simplified.

$$\frac{x^2+6x-2}{x+1}-\frac{2x+1}{x+1}+\frac{6}{x+1}=\frac{x^2+6x-2-(2x+1)+6}{x+1} \quad \text{Combine numerators.}$$

$$=\frac{x^2+6x-2-2x-1+6}{x+1} \quad \text{Distribute.}$$

$$=\frac{x^2+4x+3}{x+1} \quad \text{Simplify.}$$

$$=\frac{\overset{1}{\cancel{(x+1)}}(x+3)}{\underset{1}{\cancel{x+1}}} \quad \text{Factor.}$$

$$=x+3 \quad \text{Simplify.}$$

If the rational expressions have different denominators, first factor the denominators to see if they share any common factors. The least common denominator will include all of the common and unique factors from each rational expression. Multiply each rational expression by a form of 1 in order to obtain the common denominator. Add or subtract the resulting numerators and keep the common denominator.

Example: $\dfrac{6}{x^2 y^3} - \dfrac{4x}{yz}$

> **Solution:** Identify the least common denominator and rewrite each expression with the common denominator. Subtract the numerators.
>
> Factors in the denominators: x^2, y^3, y, z
>
> Least common denominator: $x^2 y^3 z$
>
> $\dfrac{6}{x^2 y^3} - \dfrac{4x}{yz} = \dfrac{6}{x^2 y^3} \cdot \dfrac{z}{z} - \dfrac{4x}{yz} \cdot \dfrac{x^2 y^2}{x^2 y^2}$ Rewrite with common denominator.
>
> $= \dfrac{6z}{x^2 y^3 z} - \dfrac{4x^3 y^2}{x^2 y^3 z}$ Simplify.
>
> $= \dfrac{6z - 4x^3 y^2}{x^2 y^3 z}$ Subtract.

Sometimes you will be able to factor the rational expression after you add or subtract.

Example: $\dfrac{-x-5}{x^2 - 2x - 3} + \dfrac{x}{x+1}$

> **Solution:** Factor each denominator and identify the least common denominator. Rewrite each expression with the common denominator and add the numerators. Factor the resulting rational expression to see if it can be simplified.
>
> $\dfrac{-x-5}{x^2 - 2x - 3} + \dfrac{x}{x+1} = \dfrac{-x-5}{(x-3)(x+1)} + \dfrac{x}{x+1}$ Factor the denominators.
>
> Factors in the denominators: $(x-3)$, $(x+1)$
>
> Least common denominator: $(x-3)(x+1)$

$$= \frac{-x-5}{(x-3)(x+1)} + \frac{x}{x+1} \cdot \frac{x-3}{x-3}$$

Rewrite with common denominator.

$$= \frac{-x-5}{(x-3)(x+1)} + \frac{x^2-3x}{(x-3)(x+1)}$$

Simplify.

$$= \frac{-x-5+x^2-3x}{(x-3)(x+1)}$$

Add numerators.

$$= \frac{x^2-4x-5}{(x-3)(x+1)}$$

Simplify.

$$= \frac{(x-5)\,\cancel{(x+1)}}{(x-3)\,\cancel{(x+1)}}$$

Factor numerator.

$$= \frac{(x-5)}{(x-3)}$$

Simplify.

Practice Exercises

9.7 $\quad \dfrac{x+1}{x^2 y} + \dfrac{2y}{x}$

9.8 $\quad \dfrac{x+12}{x^2-x-6} - \dfrac{x}{x+2}$

9.3 Simplify Complex Rational Expressions

A *complex rational expression* is a rational expression that contains a rational expression in the numerator, denominator, or both.

Examples of Complex Rational Expressions

$$\frac{\dfrac{1}{x+2}}{\dfrac{1}{y}} \qquad \frac{1}{3+\dfrac{x}{x^2-1}} \qquad \frac{2-\dfrac{x-1}{4}}{5} \qquad \frac{3-\dfrac{1}{x}}{4+\dfrac{1}{2x}}$$

If the numerator and denominator of a complex rational expression are single fraction expressions, one way to simplify is to treat the rational expression as a division problem. Multiply by the reciprocal of the denominator.

Example: Simplify $\dfrac{\dfrac{1}{x+2}}{\dfrac{1}{y}}$.

Solution: Write the complex rational expression as a division problem and multiply by the reciprocal.

$\dfrac{\dfrac{1}{x+2}}{\dfrac{1}{y}} = \dfrac{1}{x+2} \div \dfrac{1}{y}$ Write as a division problem.

$\qquad = \dfrac{1}{x+2} \cdot \dfrac{y}{1}$ Multiply by the reciprocal.

$\qquad = \dfrac{y}{x+2}$ Simplify.

If the numerator and denominator of a complex rational expression are not single fraction expressions, you must first simplify the numerator and denominator expressions into single fraction expressions before you divide.

Example: Simplify $\dfrac{3-\dfrac{1}{x}}{4+\dfrac{1}{2x}}$.

Solution: Simplify the numerator and denominator into single fraction expressions. Divide the numerator by the denominator by multiplying by the reciprocal of the denominator.

$\dfrac{3-\dfrac{1}{x}}{4+\dfrac{1}{2x}} = \dfrac{\dfrac{3x}{x}-\dfrac{1}{x}}{\dfrac{8x}{2x}+\dfrac{1}{2x}}$ Find common denominators.

$\qquad = \dfrac{\dfrac{3x-1}{x}}{\dfrac{8x+1}{2x}}$ Simplify numerator and denominator.

$\qquad = \dfrac{3x-1}{x} \div \dfrac{8x+1}{2x}$ Rewrite as a division problem.

$$= \frac{3x-1}{\cancel{x}^{\,1}} \cdot \frac{2\cancel{x}^{\,1}}{8x+1} \qquad \text{Multiply by the reciprocal.}$$

$$= \frac{2(3x-1)}{8x+1} = \frac{6x-2}{8x+1} \quad \text{Simplify.}$$

Often, the quickest solution for any complex rational expression is to multiply the rational expression by the least common denominator of the rational expressions in both the numerator and denominator. This will eliminate the rational expressions in the numerator and denominator, leaving a single rational expression.

Example: Simplify $\dfrac{3-\dfrac{1}{x}}{4+\dfrac{1}{2x}}$.

> **Solution:** Identify the least common denominator for the rational expressions in the numerator and denominator. Multiply the numerator and denominator by the least common denominator. Simplify the remaining rational expression.
>
> Factors in the denominators: x, $2x$
>
> Least common denominator: $2x$
>
> $$\frac{3-\dfrac{1}{x}}{4+\dfrac{1}{2x}} \cdot \frac{2x}{2x} = \frac{3(2x)-\left(\dfrac{1}{\cancel{x}_1}\right)2\cancel{x}^{\,1}}{4(2x)+\left(\dfrac{1}{\cancel{2x}_1}\right)\cancel{2x}^{\,1}} \qquad \text{Multiply by common denominator.}$$
>
> $$= \frac{6x-2}{8x+1} \qquad \text{Simplify.}$$

Notice that the two previous examples illustrate two different ways to solve the same complex rational expression.

Practice Exercises

9.9 $\dfrac{\dfrac{\dfrac{x}{y}}{3}}{y-1}$
9.10 $\dfrac{x+\dfrac{1}{x}}{2-\dfrac{1}{x^2}}$

9.4 Rational Equations

A *rational equation* is an equation that contains at least one rational expression. To solve a rational equation, first note any restrictions on the domain of the variable, that is, a value that would make the denominator of a rational expression equal to zero. To simplify the equation, multiply both sides of the equation by the least common denominator of the rational expressions. Make sure to check your solutions in the original equation.

Example: $\dfrac{3}{x-1}+\dfrac{2}{x}=\dfrac{23}{4x}$

> **Solution:** Identify the restricted values by finding values of x that will make a denominator equal to 0. Multiply both sides of the equation by the least common denominator. Solve the resulting equation and check the solution.
>
> $x-1=0$ when $x=1$.
>
> x and $4x=0$ when $x=0$.
>
> Restricted values: $x \neq 0, 1$
>
> Least common denominator: $4x(x-1)$
>
> $\left(4x(x-1)\right)\left(\dfrac{3}{x-1}+\dfrac{2}{x}\right)=\dfrac{23}{4x}\left(4x(x-1)\right)$ Multiply by least common denominator.
>
> $\left(4x\,(x-1)\right)\left(\dfrac{3}{x-1}\right)+\left(4x(x-1)\right)\left(\dfrac{2}{x}\right)=\dfrac{23}{4x}\left(4x(x-1)\right)$ Simplify.
>
> $12x+8x-8=23x-23$ Simplify.
>
> $20x-8=23x-23$ Simplify.
>
> $15=3x$ Solve.
>
> $5=x$; 5 is not a restricted value.   Solve.

Check:

$$\frac{3}{5-1}+\frac{2}{5}=\frac{23}{4\cdot 5}$$

$$\frac{3}{4}+\frac{2}{5}=\frac{23}{20}$$

$$\frac{15}{20}+\frac{8}{20}=\frac{23}{20}$$

$$\frac{23}{20}=\frac{23}{20}\ \checkmark$$

$x = 5$ is the solution.

If the only solution to a rational equation is a restricted value, the equation has no solution.

Example: $\dfrac{4}{x+3}-\dfrac{2}{x+2}=\dfrac{x-1}{(x+3)(x+2)}$

Solution: Note the restricted values for x. Multiply both sides of the equation by the least common denominator. Solve the resulting equation.

$x + 3 = 0$ when $x = -3$.

$x + 2 = 0$ when $x = -2$.

Restricted values: $x \neq -3, -2$

Least common denominator: $(x + 3)(x + 2)$

$$(x+3)(x+2)\left(\frac{4}{x+3}-\frac{2}{x+2}\right)=\left(\frac{x-1}{(x+3)(x+2)}\right)(x+3)(x+2)$$

$$\cancel{(x+3)}(x+2)\left(\frac{4}{\cancel{x+3}}\right)+(x+3)\cancel{(x+2)}\left(\frac{-2}{\cancel{x+2}}\right)=\left(\frac{x-1}{\cancel{(x+3)(x+2)}}\right)\cancel{(x+3)(x+2)}$$

$$4x + 8 - 2x - 6 = x - 1$$
$$2x + 2 = x - 1$$
$$x = -3$$

-3 is a restricted value, so the solution is extraneous.

The equation has no solution.

Practice Exercises

9.11 $\dfrac{x+5}{6} - 4 = \dfrac{-3}{x}$

9.12 $\dfrac{1}{x+4} - \dfrac{2}{x-2} = \dfrac{-3}{x-1}$

9.5 Graph Rational Functions

A *rational function* has the form $y = f(x) = \dfrac{p(x)}{q(x)}$ where $p(x)$ and $q(x)$ are polynomial functions and $q(x) \neq 0$. The domain of a rational function is the set of values of x that do not make $q(x)$ equal 0. In other words, the domain of a rational function is restricted to values of x that produce a nonzero denominator.

A function whose graph has no breaks or undefined range values is called *continuous*. A function that is *discontinuous* has a break in the graph.

Point Discontinuity

A discontinuous function that has only a single point where there is a break in the graph has *point discontinuity*. The point of discontinuity is typically illustrated on a graph by an open dot.

Some rational functions will simplify into functions that are easy to graph using graphing techniques you have already learned.

Example: Graph the rational function $f(x) = \dfrac{x^2 - 4}{x - 2}$.

> **Solution:** Identify any restrictions on the domain. Simplify the rational expression. Graph the resulting function over the identified domain.
>
> The domain is restricted to values for which $x - 2 \neq 0$.
>
> $x - 2 = 0$ when $x = 2$
>
> The domain of the function is all real numbers except 2.

$$f(x) = \frac{x^2 - 4}{x - 2}$$

$$= \frac{(x+2)\,\overset{1}{\cancel{(x-2)}}}{\underset{1}{\cancel{x-2}}} \qquad \text{Factor the rational expression.}$$

$$= x + 2 \qquad\qquad \text{Simplify.}$$

$f(x) = x + 2$ is a linear function. The graph is a line with a slope of 1 and a y-intercept of 2. Since the domain does not include 2, x cannot equal 2 and the point (2, 4) is excluded from the graph.

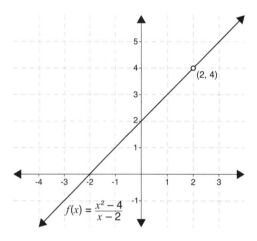

Practice Exercises

Graph each function.

9.13 $\quad f(x) = \dfrac{x^2 - x - 12}{x + 3}$ 9.14 $\quad f(x) = \dfrac{x^2 - x}{x}$

Horizontal and Vertical Asymptotes

Not all rational functions will simplify to functions that are easy to graph. Identifying asymptotes can help you graph a rational function. Recall that an asymptote is a line that a function approaches as it goes toward positive or negative infinity. Identifying the asymptotes will help divide the graph into different regions.

You can use the following rules to identify horizontal and vertical asymptotes of a rational function.

Horizontal and Vertical Asymptotes

If $y = f(x) = \dfrac{p(x)}{q(x)}$ and $p(x)$ and $q(x)$ have no common factors other

than 1 and $q(x) \neq 0$, then:

- There is a vertical asymptote at each real zero of $q(x)$.
- If the degree of $p(x)$ is less than or equal to the degree of $q(x)$, then the graph has one horizontal asymptote

  at $y = 0$ if the degree of $p(x)$ is less than the degree of $q(x)$.

  at $y = \dfrac{\text{leading coefficent of } p(x)}{\text{leading coefficient of } q(x)}$ if the degree of $p(x)$ equals the

  degree of $q(x)$.

Before evaluating a rational function for asymptotes, you must factor and simplify. If the degree of $p(x)$ is greater than the degree of $q(x)$, or the degree of the numerator is greater than the degree of the denominator, there is no horizontal asymptote.

Examples of Horizontal and Vertical Asymptotes

No Horizontal Asymptotes   **One Horizontal Asymptote**

 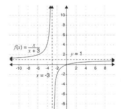 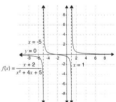

$f(x) = \dfrac{x^2}{x+2}$ $f(x) = \dfrac{x^3}{x^2-4}$ $f(x) = \dfrac{x}{x+3}$ $f(x) = \dfrac{x+2}{x^2+4x-5}$

$f(x) = \dfrac{x^3}{(x+2)(x-2)}$ $f(x) = \dfrac{x+2}{(x+5)(x-1)}$

Vertical Asymptote: Vertical Asymptotes: Vertical Asymptote: Vertical Asymptotes:
$x = -2$ $x = -2, x = 2$ $x = -3$ $x = -5, x = 1$

Horizontal Asymptote: Horizontal Asymptotes:
$y = 1$ $y = 0$

To graph a rational function, use the following steps:

- Factor $p(x)$ and $q(x)$.

- Identify the domain of the function by restricting values that make $q(x) = 0$.

- Simplify by dividing out common factors.

- Identify and graph the vertical asymptote(s) by setting $q(x) = 0$.

- Identify and graph a horizontal asymptote by comparing the degree of $p(x)$ and $q(x)$.

- Choose x values in each region of the graph, paying particular attention to how the graph behaves near any asymptote.

- Draw a smooth curve through the points in each region of the graph.

- Review any values that were restricted from the domain and indicate any point discontinuities on the graph with an open dot.

Example: Graph $f(x) = \dfrac{3}{x+1}$.

    Solution: Identify the domain and the vertical and horizontal asymptotes. Choose values in each region and evaluate the function to find ordered pairs to graph.

    If $x = -1$, then $x + 1 = 0$, so $x = -1$ is a restricted value.

    The domain of $f(x)$ is all real numbers except -1.

    There are no common factors, so the function cannot be simplified.

    If $x + 1 = 0$, then $x = -1$, so $x = -1$ is a vertical asymptote.

    $p(x) = 3$, so the degree of $p(x) = 0$.

    $q(x) = x + 1$, so the degree of $q(x) = 1$.

    The degree of $p(x) <$ the degree of $q(x)$, so $y = 0$ is a horizontal asymptote.

Choose values in each region to graph: $f(x) = \dfrac{3}{x+1}$

x	-1.25	-1.5	-2	-3	-7
$f(x)$	-12	-6	-3	-1.5	-0.5

x	-0.75	-0.5	0	1	2	5
$f(x)$	12	6	3	1.5	1	0.5

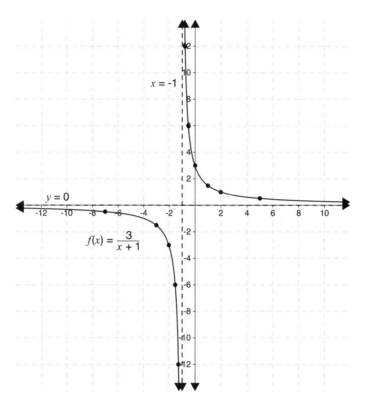

Example: Graph $f(x) = \dfrac{2x^3}{3x^3 - 27x}$.

Solution: Factor the numerator and denominator. Identify the domain. Divide out common factors and identify vertical and horizontal asymptotes. Choose values in each region and evaluate the function to find ordered pairs to graph. Review the restricted values and indicate any point discontinuities on the graph with an open dot.

$$f(x) = \frac{2x^3}{3x^3 - 27x} = \frac{2x^2 \cdot x}{3x(x^2 - 9)} = \frac{2x^2 \cdot x}{3x(x+3)(x-3)}$$

If $x = 0$, then $3x(x + 3)(x - 3) = 0$, so $x = 0$ is a restricted value.

If $x = -3$, then $3x(x + 3)(x - 3) = 0$, so $x = -3$ is a restricted value.

If $x = 3$, then $3x(x + 3)(x - 3) = 0$, so $x = 3$ is a restricted value.

The domain of $f(x)$ is all real numbers except -3, 0, and 3.

$$f(x) = \frac{2x^2 \cdot \overset{1}{\cancel{x}}}{3\underset{1}{\cancel{x}}(x+3)(x-3)} = \frac{2x^2}{3(x+3)(x-3)}$$

If $3(x + 3)(x - 3) = 0$, then $x = 3$ or -3, so $x = 3$ and $x = -3$ are vertical asymptotes.

$p(x) = 2x^2$, so the degree of $p(x) = 2$.

$q(x) = 3(x+3)(x-3) = 3x^2 - 27$, so the degree of $q(x) = 2$.

The degree of $p(x)$ = the degree of $q(x)$, so $y = \frac{2}{3}$ is a horizontal asymptote.

Choose values in each region to graph: $f(x) = \dfrac{2x^2}{3(x+3)(x-3)}$

x	-3.25	-3.5	-4	-6	-8		
$f(x)$	4.51	2.51	1.52	0.89	0.78		
x	-2.75	-2.5	-1	0	1	2.5	2.75
$f(x)$	-3.51	-1.52	-0.08	0	-0.08	-1.52	-3.51
x	3.25	3.5	4	6	8		
$f(x)$	4.51	2.51	1.52	0.89	0.78		

Recall that 0 is restricted from the domain, so there is a point discontinuity at (0, 0).

Notice that values restricted from the domain will become a vertical asymptote or points of discontinuity, depending on whether the rational expression can be simplified by factoring.

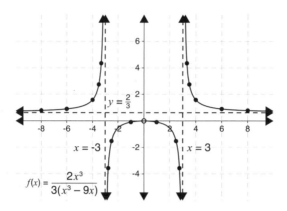

$$f(x) = \frac{2x^3}{3(x^3 - 9x)}$$

Use the original function to look for restrictions on the domain. Use the factored function to identify asymptotes.

Practice Exercises

Graph each function.

9.15 $f(x) = \dfrac{6x + 2}{x - 3}$

9.16 $f(x) = \dfrac{-2}{x - 5}$

Oblique Asymptotes

An *oblique* or *slant asymptote* is an asymptote that is neither horizontal nor vertical. An oblique asymptote occurs when the degree of the numerator is one more than the degree of the denominator.

Oblique Asymptotes

If $y = f(x) = \dfrac{p(x)}{q(x)}$, $p(x)$ and $q(x)$ have no common factors other than 1, $q(x) \neq 0$, and the degree of $p(x)$ is one more than the degree of $q(x)$, then the graph has an oblique asymptote at $y = \dfrac{p(x)}{q(x)}$ with no remainder.

The relationship between the degrees of $p(x)$ and $q(x)$ determines whether a function has a horizontal or oblique asymptote.

- If the degree of $p(x) <$ the degree of $q(x)$, there is a horizontal asymptote at $y = 0$.

- If the degree of $p(x) =$ the degree of $q(x)$, there is a horizontal asymptote at $y = \dfrac{a}{b}$ where a and b are the leading coefficients of $p(x)$ and $q(x)$.

- If the degree of $p(x)$ is one more than the degree of $q(x)$, there is an oblique asymptote at $y = \dfrac{p(x)}{q(x)}$ with no remainder.

Examples of Horizontal and Oblique Asymptotes

Horizontal Asymptotes   **Oblique Asymptote**

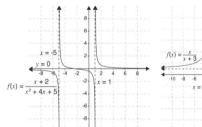
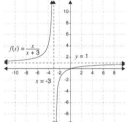
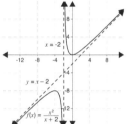

$$f(x) = \frac{x+2}{x^2+4x-5} \qquad f(x) = \frac{x}{x+3} \qquad f(x) = \frac{x^2}{x+2}$$

$$f(x) = \frac{x+2}{(x+5)(x-1)}$$

$$\begin{array}{r} x-2 \\ x+2\overline{)x^2} \\ -\left(x^2+2x\right) \\ \hline -2x \\ -(-2x-4) \\ \hline 4 \end{array}$$

Vertical Asymptotes: **Vertical Asymptote:**   **Vertical Asymptote:**
$x = -5, x = 1$ $x = -3$ $x = -2$

Horizontal Asymptote: **Horizontal Asymptote:**   **Oblique Asymptote:**
$y = 0$ $y = 1$ $y = x - 2$

Example: Graph $f(x) = \dfrac{x^3 - 1}{x^2 - 1}$.

Solution: Factor the numerator and denominator. Identify the domain. Divide out common factors and identify any vertical asymptotes and the horizontal or oblique asymptote. Choose values in each region and evaluate the function to find ordered pairs to graph. Review the restricted values and indicate any point discontinuities on the graph with an open dot.

$$f(x) = \frac{x^3 - 1}{x^2 - 1} = \frac{(x-1)(x^2 + x + 1)}{(x+1)(x-1)}$$

If $x = -1$, then $(x + 1)(x - 1) = 0$, so $x = -1$ is a restricted value.

If $x = 1$, then $(x + 1)(x - 1) = 0$, so $x = 1$ is a restricted value.

The domain of $f(x)$ is all real numbers except -1 and 1.

$$f(x) = \frac{\overset{1}{\cancel{(x-1)}}(x^2 + x + 1)}{(x+1)\,\underset{1}{\cancel{(x-1)}}} = \frac{x^2 + x + 1}{x + 1}$$

If $x + 1 = 0$, then $x = -1$, so $x = -1$ is a vertical asymptote.

$p(x) = x^2 + x + 1$, so the degree of $p(x) = 2$.

$q(x) = x + 1$ so the degree of $q(x) = 1$.

The degree of $p(x)$ is one more than the degree of $q(x)$, so the function has an oblique asymptote.

$$\begin{array}{r} x \\ x+1\overline{)x^2 + x + 1} \\ -(x^2 + x) \\ \hline 0 + 1 \end{array}$$

$y = x$ is the oblique asymptote.

Choose values in each region to graph: $f(x) = \dfrac{x^2 + x + 1}{x + 1}$

x	−1.25	−1.5	−2	−3	−5
$f(x)$	−5.25	−3.5	−3	−3.5	−5.25
x	−0.75	−0.5	0	1	3
$f(x)$	3.25	1.5	1	1.5	3.25

Recall that 1 is restricted from the domain, so there is a point discontinuity at (1, 1.5).

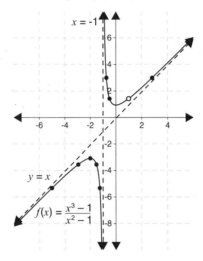

Practice Exercises

Graph each function.

9.17 $f(x) = \dfrac{2x^2}{x-3}$

9.18 $f(x) = \dfrac{x^3 + 3x^2}{x^2 + 5x + 6}$

9.6 Variation Functions

In Lesson 2.3, you were introduced to the concept of direct variation. Other variation functions include inverse variation, joint variation, and combined variation.

Direct Variation

Recall from Lesson 2.3 that a direct variation equation has the form $y = kx$, where k is the *constant of proportionality*. The phrase "y varies directly as x" or "y is directly proportional to x" indicates a direct variation function. In a direct variation function, as x increases, y increases; or, as x decreases, y decreases.

For example, if you are travelling at a certain speed, the time travelled is directly proportional to the distance travelled, or, if you are travelling at a certain speed, the longer you drive, the farther you go.

Direct variation form: $y = kx$ where $k \neq 0$ and $x \neq 0$.

If you know an ordered pair, you can easily find k, the constant of proportionality, and find the equation for the direct variation.

Example: If y varies directly as x, and $y = 6$ when $x = 2$ what is the equation of direct variation?

Solution: Substitute x and y into the direct variation equation and find k. Write the direct variation equation using the value of k.

$y = kx$ Direct variation equation.

$6 = k(2)$ Substitute known values.

$k = 3$ Divide both sides of the equation by 2.

$y = 3x$ Substitute k into the direct variation equation.

Using the direct variation equation, you can find other points that satisfy the direct variation equation.

Example: If y varies directly as x, and $y_1 = 10$ when $x_1 = 3$, what is y_2 when $x_2 = 8$?

Solution: Substitute x_1 and y_1 into the direct variation equation to find the constant of proportionality. Write the direct variation equation. Use the equation to find y when $x_2 = 8$.

$y = kx$ Direct variation equation.

$10 = k(3)$ Substitute $x_1 = 3$ and $y_1 = 10$.

$\dfrac{10}{3} = k$ Divide both sides of the equation by 3.

$y = \dfrac{10}{3}x$ Write the direct variation equation using the value of k.

$y_2 = \dfrac{10}{3}(8)$ Substitute $x_2 = 8$.

$y_2 = \dfrac{80}{3}$ Simplify.

The previous problem was solved in two steps. First, the equation of direct variation was found using the first set of values, and then the equation was used to locate the missing value in the second set of values. These steps can be combined into a single step.

For values x_1, y_1 and x_2, y_2 that satisfy the same direct variation function,

$$y_1 = kx_1 \quad \text{and} \quad y_2 = kx_2 \text{, so}$$

$$\frac{y_1}{x_1} = k \quad \text{and} \quad \frac{y_2}{x_2} = k \text{, and}$$

$$\frac{y_1}{x_1} = \frac{y_2}{x_2}$$

The previous example and other direct variation problems can be solved using the equation $\frac{y_1}{x_1} = \frac{y_2}{x_2}$.

Example: If y varies directly as x, and $y_1 = 10$ when $x_1 = 3$, what is y_2 when $x_2 = 8$?

Solution: Substitute x_1, y_1, and x_2 into $\frac{y_1}{x_1} = \frac{y_2}{x_2}$ and solve.

$\frac{y_1}{x_1} = \frac{y_2}{x_2}$ Direct variation proportion.

$\frac{10}{3} = \frac{y_2}{8}$ Substitute known values.

$\frac{80}{3} = y_2$ Multiply both sides of the equation by 8.

Practice Exercises

9.19   If y varies directly as x and $y = 7$ when $x = 14$, what is the direct variation equation?

9.20   If y varies directly as x and $y_1 = 8$ when $x_1 = 4$, what is x_2 when $y_2 = 9$?

Inverse Variation

The phrase "*y* varies inversely as *x*" or "*y* is inversely proportional to *x*" indicates an *inverse variation* function. In an inverse variation function, as *x* increases, *y* decreases; or, as *x* decreases, *y* increases.

For example, the speed you travel is inversely proportional to the time it takes to get to your destination; or, the faster you go, the less time it takes you to get there.

An inverse variation function has the form $y = \dfrac{k}{x}$, where *k* is the *constant of proportionality* and $k \neq 0$ and $x \neq 0$.

Inverse variation form: $y = \dfrac{k}{x}$ where $k \neq 0$ and $x \neq 0$.

Just as with direct variation, if you know an ordered pair, you can easily substitute the ordered pair to find *k*, the constant of proportionality, and find the equation for the inverse variation.

Example: If *y* varies inversely as *x*, and $y = 6$ when $x = 2$, what is the equation of inverse variation?

Solution: Substitute *x* and *y* into the direct variation equation and find *k*. Write the direct variation equation using the value of *k*.

$y = \dfrac{k}{x}$ Inverse variation equation.

$6 = \dfrac{k}{2}$ Substitute known values.

$12 = k$ Multiply both sides of the equation by 2.

$y = \dfrac{12}{x}$ Substitute *k* into the inverse variation equation.

Using the inverse variation equation, you can find other points that satisfy the inverse variation.

Or, if you are given one set of values and are looking for a missing value in a second set, you can use a combined approach, much like the combined approach for direct variation.

For values x_1, y_1 and x_2, y_2 that satisfy the same inverse variation function,

$$y_1 = \frac{k}{x_1} \quad \text{and} \quad y_2 = \frac{k}{x_2}, \text{ so}$$

$$x_1 y_1 = k \quad \text{and} \quad x_2 y_2 = k, \text{ and}$$

$$x_1 y_1 = x_2 y_2$$

Example: If y varies inversely as x, and $y_1 = 10$ when $x_1 = 3$, what is y_2 when $x_2 = 8$?

Solution: Substitute x_1 and y_1 into the inverse variation equation to find the constant of proportionality. Write the inverse variation equation. Use the equation to find y when $x_2 = 8$. Or, substitute x_1, y_1, and x_2 into $x_1 y_1 = x_2 y_2$ and solve.

$x_1 y_1 = x_2 y_2$	Inverse variation equation.
$3(10) = 8y_2$	Substitute known values.
$30 = 8y_2$	Simplify.
$\dfrac{30}{8} = y_2$	Divide both sides of the equation by 8.
$\dfrac{15}{4} = y_2$	Simplify.

Practice Exercises

9.21 If y varies inversely as x, and $y = 8$ when $x = 6$, what is the equation of inverse variation?

9.22 If y varies inversely as x, and $y_1 = 3$ when $x_1 = 2$, what is x_2 when $y_2 = 5$?

Joint Variation

The phrase "y varies jointly as x and z" or "y is jointly proportional to x and z" indicates a *joint variation* function. In a joint variation function, one variable varies as the product of two or more variables. For example, the amount of simple interest earned on an investment varies jointly with the interest rate and amount of time the money is invested.

A joint variation function has the form $y = kxz$, where k is the *constant of proportionality* and $k \neq 0$.

Joint variation form: $y = kxz$, where $k \neq 0$

Just as with direct and inverse variation, if you know a set of values, you can easily substitute to find k, the constant of proportionality, and find the equation for the joint variation.

Example: If y varies jointly as x and z, and $y = 6$ when $x = 2$ and $z = 2$, what is the equation of joint variation?

Solution: Substitute x, y, and z into the direct variation equation and find k. Write the direct variation equation using the value of k.

$y = kxz$	Joint variation equation.
$6 = k(2)(2)$	Substitute known values.
$1.5 = k$	Divide both sides of the equation by 4.
$y = 1.5xz$	Substitute k into the inverse variation equation.

Using the joint variation equation, you can find other values that satisfy the joint variation.

Or, if you are given a set of values and you are trying to find a missing value in another set of values, you can use a combined approach, much like the combined approach for direct or inverse variation.

For x_1, y_1, z_1 and x_2, y_2, z_2 that satisfy the same joint variation function,

$$y_1 = kx_1z_1 \quad \text{and} \quad y_2 = kx_2z_2 \text{, so}$$

$$\frac{y_1}{x_1z_1} = k \quad \text{and} \quad \frac{y_2}{x_2z_2} = k \text{, and}$$

$$\frac{y_1}{x_1z_1} = \frac{y_2}{x_2z_2}$$

Example: If y varies jointly as x and z, and $y_1 = 10$ when $x_1 = 3$ and $z_1 = 2$, what is y_2 when $x_2 = 8$ and $z_2 = 4$?

Solution: Substitute x_1, y_1, and z_1 into the joint variation equation to find the constant of proportionality. Write the joint variation equation. Use the equation to find y_2 when $x_2 = 8$ and $z_1 = 2$. Or,

substitute x_1, y_1, z_1, x_2, and z_2 into $\dfrac{y_1}{x_1 z_1} = \dfrac{y_2}{x_2 z_2}$ and solve.

$\dfrac{y_1}{x_1 z_1} = \dfrac{y_2}{x_2 z_2}$ Joint variation equation.

$\dfrac{10}{3(2)} = \dfrac{y_2}{8(4)}$ Substitute known values.

$\dfrac{10}{6} = \dfrac{y_2}{32}$ Simplify.

$32\left(\dfrac{10}{6}\right) = \left(\dfrac{y_2}{32}\right) 32$ Multiply both sides of the equation by 32.

$\dfrac{160}{3} = y_2$ Simplify.

Practice Exercises

9.23   If y varies jointly as x and z, and $y = 72$ when $x = 9$ and $z = 4$, what is the equation of joint variation?

9.24   If y varies jointly as x and z, and $y_1 = 15$ when $x_1 = 3$ and $z_1 = 4$, what is x_2 when $y_2 = 8$ and $z_2 = 2$?

Combined Variation

If a quantity varies directly or jointly and inversely with two or more quantities, it is called *combined variation*. For example, Newton's law of universal gravitation says that the gravitational force of attraction between two particles is jointly proportional to their masses and inversely proportional to the square of the distance between their centers.

The formula for Newton's law of universal gravitation is $F = G\dfrac{Mm}{d^2}$

where F is the gravitational force, M and m are the masses of the particles, d is the distance between the centers of the particles and G is the constant of proportionality.

Compare Newton's equation with the joint and inverse variation equations.

Joint variation: $y = kxz$ **Inverse variation:** $y = \dfrac{k}{x}$

Newton's equation is a combination of the joint and inverse equations. The variables that are jointly proportional are on the top of the fraction, while the variable that is inversely proportional is on the bottom, and the constant of proportionality, k, is represented by G.

To write combined variation equations, combine the necessary direct, joint, or inverse equations necessary to account for all of the relationships.

Variation Equations

- Direct variation: $y = kx$

- Joint variation: $y = kxz$

- Inverse variation: $y = \dfrac{k}{x}$

Example: y is directly proportional to p and inversely proportional to q. If $y_1 = 10$ when $p_1 = 2$ and $q_1 = 3$, what is y_2 when $p_2 = 8$ and $q_2 = 4$?

Solution: Write a combined variation equation using the direct and inverse variation equations and the variables y, p, and q. Substitute y_1, p_1, and q_1 into the combined variation equation to find the constant of proportionality. Write the combined variation equation. Use the combined variation equation to find y_2 when $p_2 = 8$ and $q_2 = 4$.

Direct variation equation: $y = kp$

Inverse variation equation: $y = \dfrac{k}{q}$

Combined variation equation: $y = \dfrac{kp}{q}$

$$y = \frac{kp}{q}$$ Combined variation equation.

$$10 = \frac{k(2)}{3}$$ Substitute known values.

$$15 = k$$ Multiply both sides of the equation by $\frac{3}{2}$.

$$y = \frac{15p}{q}$$ Write combined variation equation with $k = 15$.

$$y_2 = \frac{15(8)}{(4)}$$ Substitute $p_2 = 8$ and $q_2 = 4$.

$$y_2 = 30$$ Simplify.

Practice Exercises

y is directly proportional to m and inversely proportional to the square of n.

9.25   Write the combined variation equation if $y = 6$ when $m = 4$ and $n = 2$.

9.26   Using the equation from 9.25, what is the value of y when $m = 7$ and $n = 3$?

Chapter 10

Sequences and Series

10.1 Sequences

A *sequence* is a set of numbers in a certain pattern. Each number in the set is called a *term*. A finite sequence has a limited and finite number of terms, while an infinite sequence has an unlimited and infinite number of terms. An infinite sequence is indicated with three dots, called *ellipses*, after the last number listed to show that the sequence goes on infinitely.

{0, 2, 4, 6, 8, 10}   {0, 2, 4, 6, 8, 10, ...}

Finite sequence   Infinite sequence

The first term of a sequence is identified by a_1, the second term a_2, and so on up to the nth term, a_n. The term a_n is also called the general term. In each preceding sequence, $a_1 = 0$, $a_2 = 2$, $a_3 = 4$, $a_4 = 6$, $a_5 = 8$, and $a_6 = 10$.

Identifying Arithmetic Sequences

An *arithmetic sequence* is formed when a constant value is added to each term to find the next term. This constant value is called the *common difference*. You can find the common difference by taking a term and subtracting the preceding term.

Arithmetic Sequence

7, 10, 13, 16, 19

+3   +3   +3   +3

In the sequence above, the common difference is 3, since $10 - 7 = 3$, $13 - 10 = 3$, $16 - 13 = 3$, and $19 - 16 = 3$. You can write the common difference as the general equation $a_n - a_{n-1} = 3$.

Example: Identify whether each sequence is an arithmetic sequence. If the sequence is an arithmetic sequence, identify the common difference and find the next four terms.

     a. –3, 1, 5, 9, 13, ...

     b. 6, 8, 11, 13, 16, ...

Solution: Starting with the second term, take each term and subtract the preceding term to determine whether there is a common difference. If there is a common difference, add the common difference to the last term to find the next term. Continue adding the common difference to find the next four terms.

a. $-3, 1, 5, 9, 13, \ldots$

$1 - (-3) = 4$

$5 - 1 = 4$

$9 - 5 = 4$

$13 - 9 = 4$

The common difference is 4.

The sequence is arithmetic.

$13 + 4 = 17$

$17 + 4 = 21$

$21 + 4 = 25$

$25 + 4 = 29$

The next four terms are 17, 21, 25, and 29.

b. $6, 8, 11, 13, 16, \ldots$

$8 - 6 = 2$

$11 - 8 = 3$

There is not a common difference.

The sequence is not arithmetic.

Practice Exercises

Identify whether each sequence is an arithmetic sequence. If the sequence is an arithmetic sequence, identify the common difference and find the next four terms.

10.1 $3, 6, 9, 12, 16, 20, 24, \ldots$

10.2 $-3, -5, -7, -9, -11, \ldots$

Identifying Geometric Sequences

A *geometric sequence* is formed when a constant value is multiplied by each term to find the next term. This constant value is called the *common ratio*. You can find the common ratio by dividing a term by the preceding term.

Geometric Sequence

$$16, \ 24, \ 36, \ 54, \ 81$$

$$\times\frac{3}{2} \quad \times\frac{3}{2} \quad \times\frac{3}{2} \quad \times\frac{3}{2}$$

In the sequence above, the common ratio is $\frac{3}{2}$, since $\frac{24}{16} = \frac{3}{2}$, $\frac{36}{24} = \frac{3}{2}$, $\frac{54}{36} = \frac{3}{2}$, and $\frac{81}{54} = \frac{3}{2}$.

You can write the common ratio as the general equation $\frac{a_n}{a_{n-1}} = \frac{3}{2}$.

Example: Identify whether each sequence is a geometric sequence. If the sequence is a geometric sequence, identify the common ratio and find the next four terms.

a. 2, 10, 50, 250, 1,250 …

b. 4, 2, 1, $\dfrac{1}{2}$, $\dfrac{1}{4}$, …

Solution: Starting with the second term, divide each term by the preceding term to determine whether there is a common ratio. If there is a common ratio, multiply the last term by the common ratio to find the next term. Continue multiplying by the common ratio to find the next four terms.

a. 2, 10, 50, 250, 1,250 …

$$\frac{10}{2} = 5 \quad \frac{50}{10} = 5$$

$$\frac{250}{50} = 5 \quad \frac{1,250}{250} = 5$$

The common ratio is 5.

The sequence is geometric.

$$1,250 \cdot 5 = 6,250$$

$$6,250 \cdot 5 = 31,250$$

$$31,250 \cdot 5 = 156,250$$

$$156,250 \cdot 5 = 781,250$$

The next four terms are

6,250, 31,250, 156,250 and 781,250.

b. 4, 2, 1, $\dfrac{1}{2}$, $\dfrac{1}{4}$, …

$$\frac{2}{4} = \frac{1}{2} \quad \frac{1}{2} = \frac{1}{2}$$

$$\frac{\frac{1}{2}}{1} = \frac{1}{2} \quad \frac{\frac{1}{4}}{\frac{1}{2}} = \frac{1}{4} \cdot \frac{2}{1} = \frac{1}{2}$$

The common ratio is $\dfrac{1}{2}$.

The sequence is geometric.

$$\frac{1}{4} \cdot \frac{1}{2} = \frac{1}{8}$$

$$\frac{1}{8} \cdot \frac{1}{2} = \frac{1}{16}$$

$$\frac{1}{16} \cdot \frac{1}{2} = \frac{1}{32}$$

$$\frac{1}{32} \cdot \frac{1}{2} = \frac{1}{64}$$

The next four terms are

$$\frac{1}{8}, \frac{1}{16}, \frac{1}{32}, \frac{1}{64}.$$

Practice Exercises

Identify whether each sequence is a geometric sequence. If the sequence is a geometric sequence, identify the common ratio and find the next four terms.

10.3   2, –6, 18, –54, 162, …

10.4   3, 6, 9, 12, 15, …

Analyzing Sequences

The easiest way to determine whether a sequence is arithmetic, geometric, or neither, is to check for a common difference or a common ratio.

Example: Identify each sequence as arithmetic, geometric, or neither.

          a. 1, 3, 9, 27, 81, …

          b. 3, 5, 9, 17, 33, …

          c. 22, 26, 30, 34, 38, …

Solution: Check each sequence to see whether it has a common difference or a common ratio. A common difference indicates an arithmetic sequence; a common ratio indicates a geometric sequence. If the sequence has neither, it is not an arithmetic or a geometric sequence.

a. 1, 3, 9, 27, 81, …

Check for common difference:

$3 - 1 = 2$

$9 - 3 = 6$

There is not a common difference.

Check for common ratio:

$$\frac{3}{1} = 3 \qquad \frac{9}{3} = 3$$

$$\frac{27}{9} = 3 \qquad \frac{81}{27} = 3$$

The common ratio is 3.
The sequence is geometric.

b. 3, 5, 9, 17, 33, …

Check for common difference:

$5 - 3 = 2$

$9 - 5 = 4$

There is not a common difference.

Check for common ratio:

$$\frac{5}{3} = \frac{5}{3} \qquad \frac{9}{5} = \frac{9}{5}$$

There is not a common ratio.

The sequence is neither arithmetic nor geometric.

c. 22, 26, 30, 34, 38, …

Check for common difference:

$26 - 22 = 4$

$30 - 26 = 4$

$34 - 30 = 4$

$38 - 34 = 4$

The common difference is 4.

The sequence is arithmetic.

Practice Exercises

Identify each sequence as arithmetic, geometric, or neither.

10.5 6, 8, 11, 15, 20, …

10.6 $5, 1, \dfrac{1}{5}, \dfrac{1}{25}, \dfrac{1}{125}, \ldots$

10.7 $-9, -5, -1, 3, 7, \ldots$

10.2 Arithmetic Sequences and Series

An arithmetic sequence is formed when a constant value is added to each term to find the next term, while an arithmetic series is the sum of the terms of an arithmetic sequence.

Arithmetic Sequence

One way to analyze an arithmetic sequence is to consider it as a function $f(n)$ where the domain is the natural numbers representing the term numbers, and the range is the terms of the sequence.

Arithmetic Sequence: 0, 2, 4, 6, 8, 10, …

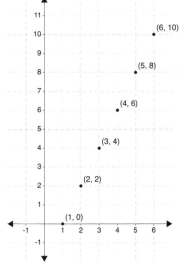

x	1	2	3	4	5
$f(x)$	0	2	4	6	8

In the graph of the arithmetic sequence, notice that the slope of a line drawn through the points would be the same as the common difference.

You can use this function approach to develop a formula to find a term in an arithmetic sequence. Consider the point-slope form.

Point-slope form: $y - y_1 = m(x - x_1)$

Let $(x_1, y_1) = (1, a_1)$ and $(x, y) = (n, a_n)$. As illustrated by the preceding graph, the slope $= m = d$. So,

$y - y_1 = m(x - x_1)$ becomes $a_n - a_1 = d(n - 1)$

$$a_n = d(n - 1) + a_1$$

The formula for the nth term in an arithmetic sequence is

$$a_n = d(n - 1) + a_1$$

where a_n is the nth term, d is the common difference, n is the term number of the nth term, and a_1 is the first term of the sequence.

Example: Find the 20th term of the arithmetic sequence –6, 1, 8, 15, 22, ….

Solution: Find the common difference by taking a term and subtracting the preceding term. Substitute the values of d, n, and a_1 into the formula for the nth term of an arithmetic sequence.

Common difference $= 1 - (-6) = 7$

$d = 7$, $n = 20$, $a_1 = -6$

$a_n = d(n - 1) + a_1$ Formula for the nth term of an arithmetic sequence.

$a_{20} = 7(20 - 1) + (-6)$ Substitute known values.

$a_{20} = 7(19) + (-6)$ Simplify.

$a_{20} = 127$ Evaluate.

The 20th term of the sequence is 127.

You can find the equation for the nth term when you are given different pieces of information.

Example: If the 8th term of an arithmetic sequence is 29 and the common difference is –2, what is the 100th term?

Solution: Substitute the known quantities in the formula for the nth term of an arithmetic sequence and solve for a_1. Use a_1, n, and d to write the equation for the 100th term. Solve to find a_{100}.

$a_8 = 29$, $n = 8$, $d = -2$

$a_n = d(n-1) + a_1$	Formula for the nth term of an arithmetic sequence.
$29 = -2(8-1) + a_1$	Substitute known values.
$29 = -14 + a_1$	Simplify.
$43 = a_1$	Add 14 to both sides of the equation.

$43 = a_1$, $d = -2$, $n = 100$

$a_n = d(n-1) + a_1$	Formula for the nth term of an arithmetic sequence.
$a_{100} = -2(100-1) + 43$	Substitute known quantities.
$a_{100} = -155$	Evaluate.

The 100th term is –155.

If you are given any two terms in the sequence, you can find other terms in the sequence. If you are given two non-consecutive terms, the terms that lie between them are called *arithmetic means*.

Example: Insert four arithmetic means between 16 and 51 and find the next four terms after 51.

Solution: If $a_1 = 16$, then the arithmetic means are a_2, a_3, a_4, and a_5, and $a_6 = 51$. Substitute a_1, a_6, and n into the formula for the nth term of an arithmetic sequence to find the value of d. Use d to find the arithmetic means and the additional terms of the sequence.

$a_1 = 16$, $a_6 = 51$, and $n = 6$

$a_n = d(n-1) + a_1$	Formula for the nth term of an arithmetic sequence.

$51 = d(6 - 1) + 16$ Substitute known values.

$51 = 5d + 16$ Simplify.

$35 = 5d$ Subtract 16 from both sides of the equation.

$7 = d$ Divide both sides of the equation by 5.

16, 23, 30, 37, 44, 51, 58, 65, 72, 79
  +7 +7 +7 +7 +7 +7 +7 +7 +7

The arithmetic means are 23, 30, 37, and 44. The next four terms in the sequence are 58, 65, 72, and 79.

Practice Exercises

10.8   Find the 100th term of the arithmetic sequence 8, 2, –4, –10, …

10.9   If the 10th term of an arithmetic sequence is 59, and the common difference is 6, what is the 50th term?

10.10   Insert three arithmetic means between 4 and –8.

Arithmetic Series

An *arithmetic series* is the sum of the terms of an arithmetic sequence. If the sequence is infinite, then the corresponding series is an *infinite series*. If the sequence is finite, then the corresponding series is a *finite series* or a *partial sum*. The symbol S_n indicates the partial sum of n terms.

$$S_n = a_1 + a_2 + a_3 + \ldots + a_n$$

You will find two formulas useful in calculating partial sums. The formula you use depends on the information you are given.

Known Values	Formula
a_1, a_n, n	$S_n = \dfrac{n}{2}(a_1 + a_n)$
a_1, n, d	$S_n = \dfrac{n}{2}\left[2a_1 + (n-1)d\right]$

where a_1 is the first term of the sequence, a_n is the nth term in the sequence, n is the term number of the nth term, and d is the common difference.

Example: Find the sum of the first five terms of the arithmetic sequence $-1, -4, -7, -10, -13$.

    Solution: Identify the known quantities, choose one of the partial sum formulas, and evaluate.

$a_1 = -1, a_5 = -13, n = 5$.

$S_n = \dfrac{n}{2}(a_1 + a_n)$ Partial sum formula.

$S_5 = \dfrac{5}{2}(-1 + (-13))$ Substitute known values.

$S_5 = \dfrac{5}{2}(-14)$ Simplify.

$S_5 = -35$ Evaluate.

If you don't have all the values you need to use one of the partial sum formulas, you can find additional values by using the equation for the nth term of an arithmetic series.

$$a_n = d(n-1) + a_1$$

where a_n is the nth term, d is the common difference, n is the term number of the nth term.

Example: Find the value of S_{20} if $a_6 = 10$ and $a_7 = 13$.

    Solution: Find the common difference by subtracting a_6 from a_7. Use the formula for the nth term of an arithmetic series to find the value of a_1. Substitute a_1, n, and d into the second partial sum formula and evaluate.

$d = a_7 - a_6 = 13 - 10 = 3$ Find common difference.

$a_7 = 13, n = 7, d = 3$ Known values.

$a_n = d(n-1) + a_1$ Formula for nth term of an arithmetic sequence.

$13 = 3(7-1) + a_1$ Substitute known values.

$13 = 18 + a_1$ Simplify.

$-5 = a_1$ Subtract 18 from both sides of the equation.

$a_1 = -5, n = 20, d = 3$ Known values.

$S_n = \dfrac{n}{2}\left[2a_1 + (n-1)d\right]$ Partial sum formula.

$S_{20} = \dfrac{20}{2}\left[2(-5) + (20-1)3\right]$ Substitute known values.

$S_{20} = 10\left[-10 + 57\right]$ Simplify.

$S_{20} = 470$ Evaluate.

Practice Exercises

10.11   Find the sum of the first 10 terms of the arithmetic sequence –10, –6, –2, 2, 6, …

10.12   Find the value of S_{50} if $a_4 = 6$ and $a_9 = 16$.

10.3 Geometric Sequences and Series

A geometric sequence is formed when a constant value is multiplied to each term to find the next term, while a geometric series is the sum of the terms of a geometric sequence.

Geometric Sequence

One way to analyze a geometric sequence is to consider it as a function $f(n)$ where the domain is the natural numbers representing the term numbers, and the range is the terms of the sequence.

Geometric Sequence: 2, 4, 8, 16, 32 …

x	1	2	3	4	5
$f(x)$	2	4	8	16	32

The graph of the geometric sequence is an exponential function with base r.

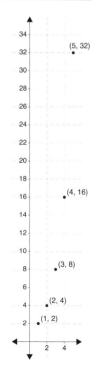

The formula for the nth term in a geometric sequence is

$$a_n = a_1 r^{n-1}$$

where a_1 is the first term of the sequence and r is the common ratio.

Example: Find the 10th term of the sequence 243, 81, 27, 9, ….

Solution: Find the common ratio by dividing a term by the preceding term. Substitute the values of r, a_1, and n into the formula for the nth term of a geometric sequence.

Common ratio $= \dfrac{81}{243} = \dfrac{1}{3}$

$r = \dfrac{1}{3}$, $a_1 = 243$, $n = 10$

$a_n = a_1 r^{n-1}$ Formula for the nth term of a geometric sequence.

$a_{10} = 243\left(\dfrac{1}{3}\right)^{10-1}$ Substitute known values.

$a_{10} = 3^5\left(\dfrac{1}{3}\right)^9$ Simplify.

$a_{10} = \dfrac{3^5}{3^9}$ Simplify.

$a_{10} = \dfrac{1}{3^4} = \dfrac{1}{81}$ Evaluate.

The 10th term of the sequence is $\dfrac{1}{81}$.

Example: If the 4th term of a geometric sequence is 54 and the common ratio is -3, what is the 15th term?

Solution: Substitute the known quantities in the formula for the nth term of a geometric sequence and solve for a_1. Use a_1, r, and n to find the 15th term.

$a_4 = 54$, $n = 4$, $r = -3$

$a_n = a_1 r^{n-1}$ Formula for the nth term of a geometric sequence.

$54 = a_1(-3)^{4-1}$ Substitute known values.

$54 = a_1(-3)^3$ Simplify.

$54 = -27a_1$ Simplify.

$-2 = a_1$ Divide both sides of the equation by -27.

$a_1 = -2, r = -3, n = 15$

$a_n = a_1 r^{n-1}$ Formula for the nth term of a geometric sequence.

$a_{15} = -2(-3)^{15-1}$ Substitute known quantities.

$a_{15} = -2(-3)^{14}$ Simplify.

$a_{15} = -9,565,938$ Evaluate.

The 15th term is $-9,565,938$.

If you are given any two terms in the sequence, you can find other terms in the sequence. If you are given two non-consecutive terms, the terms that lie between them are called *geometric means*.

Example: Insert three geometric means between 3 and 768.

Solution: If $a_1 = 3$, the arithmetic means are a_2, a_3, and a_4. $a_5 = 768$. Substitute a_1, a_5, and n into the formula for the nth term of a geometric sequence to find the value of r. Use r to find the geometric means.

$a_1 = 3$, $a_5 = 768$, and $n = 5$

$a_n = a_1 r^{n-1}$ Formula for the nth term of an arithmetic sequence.

$768 = 3r^{5-1}$ Substitute known values.

$768 = 3r^4$ Simplify.

$256 = r^4$ Divide both sides of the equation by 3.

$\pm\sqrt[4]{256} = r$ Take the fourth root of both sides of the equation.

$\pm 4 = r$ Evaluate.

A common ratio of 4 or –4 will produce a sequence that has three geometric means between 3 and 768.

$$\underbrace{3,}_{} \underbrace{12,}_{\times 4} \underbrace{48,}_{\times 4} \underbrace{19}_{\times 4}\underbrace{2, 768}_{\times 4}$$

$$\underbrace{3,}_{} \underbrace{-12,}_{\times(-4)} \underbrace{48,}_{\times(-4)} \underbrace{-19}_{(\times -4)}\underbrace{2, 768}_{(\times -4)}$$

The geometric means are 12, 48, and 192 or –12, 48, and –192.

Practice Exercises

10.13   Find the 12th term of the geometric sequence 2,048, 1,024, 512, 256, …

10.14   If the 6th term of a geometric sequence is 256 and the common ratio is –2, what is the 20th term?

10.15   Insert four geometric means between 5 and 1,215.

Geometric Series

A *geometric series* is the sum of the terms of a geometric sequence. If the sequence is infinite, then the corresponding series is an *infinite series*. If the sequence is finite, then the corresponding series is a *finite series* or a *partial sum*. The symbol S_n indicates the partial sum of n terms.

$$S_n = a_1 + a_2 + a_3 + ... + a_n$$

$$S_n = a_1 + a_1 r + a_1 r^2 + ... + a_1 r^{n-1}$$

You will find two formulas useful in calculating partial sums. The formula you use depends on the information you are given.

Known Values	Formula
a_1, a_n, r	$S_n = \dfrac{a_1 - a_n r}{1 - r}, r \neq 1$
a_1, n, r	$S_n = \dfrac{a_1 - a_1 r^n}{1 - r}, r \neq 1$

where a_1 is the first term of the sequence, a_n is the nth term in the sequence, n is the term number of the nth term, and r is the common ratio.

Example: Find the sum of the first five terms of a geometric sequence with a first term of -1 and a common ratio of 4.

Solution: Identify the known quantities, choose one of the partial sum formulas, and evaluate.

$a_1 = -1, n = 5, r = 4$

$S_n = \dfrac{a_1 - a_1 r^n}{1 - r}$ Partial sum formula.

$S_5 = \dfrac{-1 - (-1)4^5}{1 - 4}$ Substitute known values.

$S_5 = \dfrac{-1 + 1,024}{-3}$ Simplify.

$S_5 = \dfrac{1,023}{-3}$ Simplify.

$S_5 = -341$ Evaluate.

If you don't have all the values you need to use one of the partial sum formulas, you can find additional values by using the formula for the nth term of a geometric series.

$$a_n = a_1 r^{n-1}$$

where a_n is the nth term, r is the common ratio, and n is the term number.

Example: Find the value of S_6 if $a_1 = 1,024$ and $a_6 = 32$.

Solution: Find r by using the formula for the nth term of a geometric series. Substitute a_1, a_6, n, and r into the first partial sum formula and evaluate.

$a_n = a_1 r^{n-1}$ Formula for the nth term of a geometric sequence.

$32 = 1,024 r^{6-1}$ Substitute known values.

$$32 = 1{,}024r^5 \quad \text{Simplify.}$$

$$\frac{32}{1{,}024} = r^5 \qquad \text{Divide both sides of the equation by 1,024.}$$

$$\frac{1}{32} = r^5 \qquad \text{Simplify.}$$

$$\left(\frac{1}{2}\right)^5 = r^5 \qquad \text{Rewrite } \frac{1}{32} \text{ as a fraction to the fifth power.}$$

$$\frac{1}{2} = r \qquad \text{Solve for } r.$$

$$a_1 = 1{,}024,\ a_6 = 32,\ n = 6,\ r = \frac{1}{2} \qquad \text{Known values.}$$

$$S_n = \frac{a_1 - a_n r}{1 - r} \qquad \text{Formula for the partial sum.}$$

$$S_6 = \frac{1{,}024 - 32\left(\dfrac{1}{2}\right)}{1 - \dfrac{1}{2}} \qquad \text{Substitute known values.}$$

$$S_6 = \frac{1{,}024 - 16}{\dfrac{1}{2}} \qquad \text{Simplify.}$$

$$S_6 = \frac{1{,}008}{\dfrac{1}{2}} \qquad \text{Simplify.}$$

$$S_6 = \frac{1{,}008}{1} \cdot \frac{2}{1} \qquad \text{Simplify.}$$

$$S_6 = 2{,}016 \qquad \text{Evaluate.}$$

Practice Exercises

10.16   Find S_8 if $a_1 = -2$ and $r = \dfrac{1}{3}$.

10.17   Find S_7 if $a_1 = 96$ and $a_6 = 3{,}072$.

10.4 Summation Notation

A series may be written using *sigma notation* or *summation notation*. This notation utilizes the uppercase Greek letter sigma, Σ. Below the sigma, the *index of summation*, or the variable, is noted. The upper and lower limits of the index are noted above and below the sigma. The lower limit indicates the smallest number to be substituted for the variable, while the upper limit is the largest number to be substituted for the variable. The replacements for the variable are always consecutive integers.

$$\sum_{k=1}^{4} 3k + 2 \qquad \text{The summation of } 3k + 2 \text{ as } k \text{ goes from 1 to 4.}$$

The expression that follows the sigma is the expression for the general term of the series. So, for the summation above, $a_k = 3k + 2$.

$$\sum_{k=1}^{4} 3k + 2 = [3(1) + 2] + [3(2) + 2] + [3(3) + 2] + [3(4) + 2]$$
$$= 5 + 8 + 11 + 14$$
$$= 38$$

Example: Write out the indicated series and find the sum.

$$\sum_{j=3}^{6} 2^j$$

Solution: Substitute 3, 4, 5, and 6 for j and add.

$$\sum_{j=3}^{6} 2^j = 2^3 + 2^4 + 2^5 + 2^6$$
$$= 8 + 16 + 32 + 64$$
$$= 120$$

Instead of writing out the series and adding to find the sum, you can use the partial sum formulas.

Series	Formula
Arithmetic	$S_n = \dfrac{n}{2}(a_1 + a_n)$ or $S_n = \dfrac{n}{2}\left[2a_1 + (n-1)d\right]$
Geometric	$S_n = \dfrac{a_1 - a_n r}{1 - r}$ or $S_n = \dfrac{a_1 - a_1 r^n}{1 - r}$

Example: Evaluate $\displaystyle\sum_{n=1}^{10} 2n-1$

Solution: The summation represents an arithmetic series with a common difference of 2. Since the index goes from 1 to 10, $n = 10$. Evaluate the expression for $n = 1$ and $n = 10$ to identify a_1 and a_{10}.

Substitute n, a_1, and a_{10} into $S_n = \dfrac{n}{2}(a_1 + a_n)$.

When $n = 1$, $2n - 1 = 2(1) - 1 = 1 = a_1$.

When $n = 10$, $2n - 1 = 2(10) - 1 = 19 = a_{10}$.

$S_n = \dfrac{n}{2}(a_1 + a_n)$ Partial sum formula.

$S_{10} = \dfrac{10}{2}(1 + 19)$ Substitute known values.

$S_{10} = 100$ Evaluate.

Notice that for the previous example you could have used the partial sum formula $S_n = \dfrac{n}{2}\left[2a_1 + (n-1)d\right]$ for $n = 10$, $a_1 = 1$ and $d = 2$.

Watch out for summations where the lower limit of the index is not 1. Identify a_1, and a_n if needed, using the lower and upper limits. Identify n, the number of terms, by subtracting the lower limit from the upper limit and adding 1.

Example: Evaluate $\displaystyle\sum_{k=4}^{9} 3^k$

Solution: The summation is a geometric series with a common ratio of 3. The lower limit is 4 and the upper limit is 9, so there are $9 - 4 + 1 = 6$ terms.

Evaluate the expression for $k = 4$ to identify a_1. Substitute n, a_1, and r into $S_n = \dfrac{a_1 - a_1 r^n}{1 - r}$.

When $k = 4$, $3^k = 3^4 = 81 = a_1$.

$$S_n = \frac{a_1 - a_1 r^n}{1 - r}$$ Partial sum formula.

$$S_6 = \frac{81 - 81(3)^6}{1 - 3}$$ Substitute known values.

$$S_6 = \frac{81 - 59{,}049}{1 - 3}$$ Simplify.

$$S_6 = 29{,}484$$ Evaluate.

You can write a geometric or an arithmetic series using summation notation by finding the expression for a_n.

Series   **Formula**

Arithmetic $a_n = d(n-1) + a_1$

Geometric $a_n = a_1 r^{n-1}$

Example: Write the series $2 + 6 + 18 + 54 + 162$ in summation notation.

Solution: The series has five terms, so if the lower bound is 1, then the upper bound is 5. Identify the common difference or common ratio and substitute the common difference or common ratio and a_1 into the appropriate formula for a_n. Substitute the value of a_n in the summation notation and write the lower and upper bounds.

$6 - 2 = 4$ $\dfrac{6}{2} = 3$ $\dfrac{18}{6} = 3$

$18 - 6 = 12$

There is not a common difference. $\dfrac{54}{18} = 3$ $\dfrac{162}{54} = 3$

The common ratio is 3.

The series is a geometric series with five terms and a common ratio of 3.

$a_1 = 2, \; r = 3$

$a_n = a_1 r^{n-1}$ Formula for the nth term of a geometric sequence.

$a_n = 2(3^{n-1})$ Substitute known values.

$$\sum_{n=1}^{5} 2(3^{n-1})$$ Write in summation notation with upper and lower limits.

Notice that in the previous example the summation could also be written as $\sum_{n=0}^{4} 2(3^n)$. When $n = 0$, $2(3^0) = 2$, and 2 is the first term of the series.

Practice Exercises

10.18   Write out the series and find the sum for $\sum_{m=0}^{5} -6m$

10.19   Evaluate $\sum_{k=1}^{6} 2\left(\frac{1}{2}\right)^{k-1}$ using a partial sum formula.

10.20   Evaluate $\sum_{t=7}^{12} -2t + 6$ using a partial sum formula.

10.21   Write the series $5 + 8 + 11 + 14 + 17$ in sigma notation.

10.5 Infinite Geometric Series

The partial sums of an arithmetic series and some geometric series will increase without limit, or *diverge*, as the number of terms increases.

A geometric series will *converge*, or approach a finite sum, if $|r| < 1$. If $|r| \geq 1$, the series diverges and has no sum.

Divergent and Convergent Geometric Series

Divergent	Convergent
$a_n = 2^n$	$a_n = \left(\frac{1}{2}\right)^n$
$|r| \geq 1$	$|r| < 1$
$S_n = 2 + 4 + 8 + 16 + \ldots + 2^n$	$S_n = \frac{1}{2} + \frac{1}{4} + \frac{1}{8} + \frac{1}{16} + \ldots + \left(\frac{1}{2}\right)^n$

The sums increase without limit.   The sums approach a finite value.

Notice how the graph of the partial sums of the geometric series represented by $a_n = \left(\dfrac{1}{2}\right)^n$ appears to be approaching 1. The infinite series $\dfrac{1}{2} + \dfrac{1}{4} + \dfrac{1}{8} + \dfrac{1}{16} + \ldots$ is said to converge at 1 or approach 1 as a limit.

Example: Determine whether each geometric series converges or diverges.

   a. $1 - 0.1 + 0.01 - 0.001 + \ldots$

   b. $1 + 3 + 9 + 27 + \ldots$

  Solution: Determine the common ratio for each series. If $|r| < 1$, the series converges. If $|r| \geq 1$, the series diverges.

  a. $r = \dfrac{-0.1}{1} = -0.1$ b. $r = \dfrac{3}{1} = 3$

   $|r| = 0.1 < 1$ $|r| = 3 \geq 1$

  The series converges.   The series diverges.

If a series converges, you can find the sum. Consider the partial sum formula $S_n = \dfrac{a_1 - a_1 r^n}{1 - r}$. As n increases, if $|r| < 1$, r^n approaches 0.

Therefore, $a_1 - a_1 r^n$ approaches $a_1 - a_1(0)$ or a_1. Therefore, the sum approaches $\dfrac{a_1}{1 - r}$.

The sum of a convergent infinite geometric series where $|r| < 1$ is

$$S = \dfrac{a_1}{1 - r}.$$

Example: Find the sum of each geometric series, if it exists.

   a. $2 - 6 + 18 - 54 + \ldots$

   b. $12 + 4 + \dfrac{4}{3} + \dfrac{4}{9} + \ldots$

Solution:

a. $r = \dfrac{-6}{2} = -3$　　　　　　b. $r = \dfrac{4}{12} = \dfrac{1}{3}$

$|r| = 3 \geq 1$　　　　　　　　　$|r| = \dfrac{1}{3} < 1$

The series diverges
and has no sum.

The series converges
and has a sum.

$a_1 = 12, \; r = \dfrac{1}{3}$

$S = \dfrac{a_1}{1-r}$

$S = \dfrac{12}{1 - \dfrac{1}{3}} = \dfrac{12}{\dfrac{2}{3}} = 12 \cdot \dfrac{3}{2} = 18$

The sum of the series is 18.

Summation notation can be used to express an infinite series. The infinity symbol, ∞, is used to represent the upper limit of the index.

Example: Find the sum of the series $\displaystyle\sum_{n=1}^{\infty} 5\left(\dfrac{1}{4}\right)^n$, if it exists.

Solution: Identify the value of r and verify that the series is convergent. If so, calculate the value of a_1. Substitute a_1 and r into

$S = \dfrac{a_1}{1-r}$ to find the sum.

$r = \dfrac{1}{4}$

$|r| = \dfrac{1}{4} < 1$, so the series converges.

$a_1 = 5\left(\dfrac{1}{4}\right)^1 = \dfrac{5}{4}$　　　Substitute $n = 1$ to find a_1.

$$S = \frac{a_1}{1-r}$$ Convergent sum formula.

$$S = \frac{\dfrac{5}{4}}{1 - \dfrac{1}{4}} = \frac{\dfrac{5}{4}}{\dfrac{3}{4}} = \frac{5}{4} \cdot \frac{4}{3} = \frac{5}{3}$$ Substitute and simplify.

The sum of the series is $\dfrac{5}{3}$.

Practice Exercises

10.22 Determine whether each geometric series converges or diverges.

   a. 1, 5, 25, 125, …

   b. 81, 54, 27, 9, …

10.23 Find the sum of each geometric series, if it exists.

   a. $6 + 12 + 24 + 48 + \ldots$

   b. $-18 + -9 + \dfrac{-9}{2} + \dfrac{-9}{4} + \ldots$

10.24 Find the sum of $\displaystyle\sum_{j=3}^{\infty} 6\left(\frac{2}{3}\right)^j$.

10.6 The Binomial Theorem

Pascal's triangle is the special triangle pattern shown below.

```
                    1
                 1     1
              1     2     1
           1     3     3     1
        1     4     6     4     1
     1     5    10    10     5     1
```

There are many different patterns in the table.

Example: What are the numbers in the next row of Pascal's Triangle?

> **Solution:** Each row has one more term than the preceding row, so the next row will have seven entries. The first and last terms are 1. The second term is formed by adding the first and second terms of the preceding row. The third term is formed by adding the second and third terms of the preceding row. The remaining terms are found in the same way.

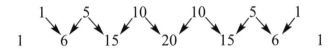

Pascal's triangle is related to the expansions of simple binomials.

$$(a+b)^0 = \qquad\qquad 1$$

$$(a+b)^1 = \qquad\qquad a + b$$

$$(a+b)^2 = \qquad\qquad a^2 + 2ab + b^2$$

$$(a+b)^3 = \qquad\qquad a^3 + 3a^2b + 3ab^2 + b^3$$

$$(a+b)^4 = \qquad a^4 + 4a^3b + 6a^2b^2 + 4ab^3 + b^4$$

$$(a+b)^5 = a^5 + 5a^4b + 10a^3b^2 + 10a^2b^3 + 5ab^4 + b^5$$

Notice the patterns in each binomial expansion.

- The coefficients are the same as the terms in row $n + 1$ of Pascal's triangle.

- There are $n + 1$ terms.

- The first term is a^n and the last term is b^n.

- In each term after the first, the exponents of a decrease by 1 and the exponents of b increase by 1.

- The sum of the exponents in each term is n.

You can use Pascal's triangle to expand binomials.

Example: Expand $(a+b)^6$.

>**Solution:** The expansion will have the coefficients from the seventh row of Pascal's triangle, since $n + 1 = 6 + 1 = 7$. The previous example lists the terms of the seventh row of Pascal's triangle. Use the coefficients and the patterns identified in the preceding binomial expansion to expand the binomial.

Seventh row of Pascal's triangle: 1, 6, 15, 20, 15, 6, 1

$$(a+b)^6 = a^6 + 6a^5b + 15a^4b^2 + 20a^3b^3 + 15a^2b^4 + 6ab^5 + b^6$$

While Pascal's triangle is useful for smaller values of n, calculating the rows necessary for larger values of n is inefficient. Factorials can be used to calculate the terms in Pascal's triangle and the coefficients of a binomial expansion.

A *factorial* is the product of a number and all of the natural numbers less than that number. Recall that the natural numbers are 1, 2, 3, 4, 5, ...

>**Factorial**
>$n! = n(n-1)(n-2) \ldots (1)$
>$1! = 1$
>$2! = 2 \cdot 1$
>$3! = 3 \cdot 2 \cdot 1$

Note that by definition $0! = 1$.

Example: Find 5!

>**Solution:** Multiply 5 by the natural numbers less than it.
>$5! = 5(4)(3)(2)(1) = 120$

The coefficient of any term in the binomial expansion $(a+b)^n$ is

$$\frac{n!}{(\text{exponent of } a)!(\text{exponent of } b)!}$$

Example: Find the fourth term of $(a+b)^{12}$.

>**Solution:** In the first term, the exponent of a is 12. In the second term, the exponent of a is 10. In the third term, the exponent of a is 10. In the fourth term, the exponent of a is 9 and the exponent of b is $12 - 9 = 3$. Substitute $a = 9$ and $b = 3$ into the formula for the coefficient of any term in a binomial expansion.

$$\frac{n!}{(\text{exponent of } a)!(\text{exponent of } b)!}$$

$$= \frac{12!}{9!3!}$$ Substitute known values.

$$= \frac{\overbrace{12 \cdot 11 \cdot 10 \cdot 9 \cdot 8 \cdot 7 \cdot 6 \cdot 5 \cdot 4 \cdot 3 \cdot 2 \cdot 1}^{12!}}{\underbrace{(9 \cdot 8 \cdot 7 \cdot 6 \cdot 5 \cdot 4 \cdot 3 \cdot 2 \cdot 1)}_{9!}\underbrace{(3 \cdot 2 \cdot 1)}_{3!}}$$ Write out factorials.

$$= \frac{\overset{2}{\cancel{12}} \cdot 11 \cdot 10 \cdot \cancel{9} \cdot \cancel{8} \cdot \cancel{7} \cdot \cancel{6} \cdot \cancel{5} \cdot \cancel{4} \cdot \cancel{3} \cdot \cancel{2} \cdot 1}{(\cancel{9} \cdot \cancel{8} \cdot \cancel{7} \cdot \cancel{6} \cdot \cancel{5} \cdot \cancel{4} \cdot \cancel{3} \cdot \cancel{2} \cdot 1)(\cancel{3} \cdot \cancel{2} \cdot 1)}$$ Identify common factors.

$$= 2 \cdot 11 \cdot 10$$ Simplify.

$$= 220$$ Evaluate.

The fourth term is $220a^9b^3$.

Notice that many of the factors in the product cancel out so that you don't have to multiply all the numbers.

The Binomial Theorem pulls together all these patterns and expresses how to expand a binomial.

The Binomial Theorem

If n is a positive integer, then

$$(a+b)^n = a^n + \frac{n}{1}a^{n-1}b + \frac{n(n-1)}{2 \cdot 1}a^{n-2}b^2 + \frac{n(n-1)(n-2)}{3 \cdot 2 \cdot 1}a^{n-3}b^3 + \ldots + b^n$$

The Binomial Theorem can also be written in factorial and sigma notation.

The Binomial Theorem: Factorial Notation

$$(a+b)^n = a^n + \frac{n!}{(n-1)!1!}a^{n-1}b + \frac{n!}{(n-2)!2!}a^{n-2}b^2 + \frac{n!}{(n-3)!3!}a^{n-3}b^3 + \ldots + b^n$$

The Binomial Theorem: Sigma Notation

$$(a+b)^n = \sum_{r=0}^{n} \frac{n!}{(n-r)!r!}a^{n-r}b^r$$

Example: Expand $(x + y)^5$.

Solution: Use the Binomial Theorem with $a = x$ and $b = y$.

$$(a+b)^n = a^n + \frac{n}{1}a^{n-1}b + \frac{n(n-1)}{2\cdot1}a^{n-2}b^2 + \frac{n(n-1)(n-2)}{3\cdot2\cdot1}a^{n-3}b^3 + ... + b^n$$

$$(x+y)^5 = x^5 + \frac{5}{1}x^4y + \frac{5(4)}{2\cdot1}x^3y^2 + \frac{5(4)(3)}{3\cdot2\cdot1}x^2y^3 + \frac{5(4)(3)(2)}{4\cdot3\cdot2\cdot1}xy^4 + y^5$$

$$(x+y)^5 = x^5 + 5x^4y + 10x^3y^2 + 10x^2y^3 + 5xy^4 + y^5$$

Notice that the coefficients of the binomial expansion are equal to the sixth row of Pascal's triangle.

If the coefficients of the binomial are not equal to one, the coefficients will not be symmetric and will not match the terms in a row of Pascal's triangle.

Example: Expand $(2x - 3)^4$.

Solution: Use the Binomial Theorem with $a = 2x$ and $b = -3$.

$$(a+b)^n = a^n + \frac{n}{1}a^{n-1}b + \frac{n(n-1)}{2\cdot1}a^{n-2}b^2 + \frac{n(n-1)(n-2)}{3\cdot2\cdot1}a^{n-3}b^3 + ... + b^n$$

$$(2x-3)^4 = (2x)^4 + \frac{4}{1}(2x)^3(-3) + \frac{4(3)}{2\cdot1}(2x)^2(-3)^2 + \frac{4(3)(2)}{3\cdot2\cdot1}(2x)(-3)^3 + (-3)^4$$

$$= 16x^4 + 4(8)x^3(-3) + 6(4)x^29 + 4(2x)(-27) + 81$$

$$= 16x^4 - 96x^3 + 216x^2 - 216x + 81$$

If you are only looking for a single term of a binomial expansion, the sigma notation form is especially helpful.

Example: Find the fifth term of $(x - 4)^8$.

Solution: Write the expansion in sigma notation. Since r starts at 0, use $r = 4$ for the fifth term.

$$(a+b)^n = \sum_{r=0}^{n} \frac{n!}{(n-r)!r!}a^{n-r}b^r$$

$$(x-4)^8 = \sum_{r=0}^{n} \frac{8!}{(8-r)!r!} x^{8-r}(-4)^r$$ Substitute known values.

If $r = 4$, the fifth term $= \dfrac{8!}{(8-4)!4!} x^{8-4}(-4)^4$ Evaluate for $r = 4$.

$= \dfrac{8!}{4!4!} x^4(256)$ Simplify.

$= \dfrac{\overset{2}{\cancel{8}} \cdot 7 \cdot \cancel{6} \cdot 5 \cdot \cancel{4} \cdot \cancel{3} \cdot \cancel{2} \cdot 1}{(\cancel{4} \cdot \cancel{3} \cdot \cancel{2} \cdot 1)(\cancel{4} \cdot \cancel{3} \cdot \cancel{2} \cdot 1)} x^4(256)$ Write out factorials.

$= 70x^4(256)$ Simplify.

$= 17{,}920x^4$ Simplify.

The fifth term is $17{,}920x^4$.

Practice Exercises

10.25   Write out the eighth and ninth rows of Pascal's triangle.

10.26   Find 7!

10.27   What is the seventh term of $(a+b)^{12}$?

10.28   Expand $(x+y)^6$.

10.29   Expand $(3x+1)^4$.

Appendix

Answers to Exercises

Chapter 1

1.1 $6 \div 2(1+2) = 6 \div 2(3) = 3(3) = 9$; Note that if you use the standard convention taught in Chapter 1 and multiply and divide in order from left to right, the answer is 9. If you multiply and then divide, you would get 1 because $6 \div 2(1+2) = 6 \div 2(3) = 6 \div 6 = 1$.

1.2 $12 \div 4 + 2 \cdot 6 - (4-3)^3 = 12 \div 4 + 2 \cdot 6 - (1)^3 = 12 \div 4 + 2 \cdot 6 - 1 = 3 + 12 - 1 = 14$

1.3 After multiplying the number by 1, the result is the same number; Identity Property of Multiplication

1.4 The order of the addends is reversed; Commutative Property of Addition

1.5 The order of the numbers being multiplied is reversed; Commutative Property of Multiplication

1.6 $8(20 + 3) = 8(20) + 8(3) = 160 + 24 = 184$

1.7 $9(14) + 9(16) = 9(14 + 16) = 9(30) = 270$

1.8 4 was subtracted from both sides; Subtraction Property of Equality

1.9 Both sides were multiplied by $\frac{3}{2}$; Multiplication Property of Equality

1.10 $5x - 3 = 6 + 3x$; $2x - 3 = 6$; $2x = 9$; $x = \frac{9}{2} = 4.5$

1.11 $\frac{2}{5}(4x-1) = -\frac{3}{10}(4-3x)$; $\overset{2}{\cancel{10}}\left(\frac{2}{\cancel{5}_1}(4x-1)\right) = \overset{1}{\cancel{10}}\left(-\frac{3}{\cancel{10}_1}(4-3x)\right)$;

$4(4x-1) = -3(4-3x)$; $16x - 4 = -12 + 9x$; $7x - 4 = -12$; $7x = -8$;

$x = -\frac{8}{7}$

1.12 $R = DL + S$; $R - S = DL$; $\frac{R-S}{L} = D$; $L \neq 0$

1.13 $3x + 4y = 12$; $4y = -3x + 12$; $y = \dfrac{-3x + 12}{4}$; $y = -\dfrac{3}{4}x + 3$

1.14 $|3x - 2| = 4$; $3x - 2 = 4$ or $3x - 2 = -4$; If $3x - 2 = 4$; $3x = 6$;

$x = 2$; If $3x - 2 = -4$; $3x = -2$; $x = -\dfrac{2}{3}$;

Check $x = 2$: $|3(2) - 2| = 4$; $|4| = 4$ ✔

Check $x = -\dfrac{2}{3}$: $\left|3\left(-\dfrac{2}{3}\right) - 2\right| = 4$; $|-4| = 4$ ✔; $x = 2, -\dfrac{2}{3}$

1.15 $3|2x + 7| = 3x - 6$; $|2x + 7| = x - 2$; $2x + 7 = x - 2$ or

$2x + 7 = -(x - 2)$; If $2x + 7 = x - 2$; $x + 7 = -2$; $x = -9$;

If $2x + 7 = -(x - 2)$; $2x + 7 = -x + 2$; $3x + 7 = 2$; $3x = -5$; $x = -\dfrac{5}{3}$

Check $x = -9$: $3|2(-9) + 7| = 3(-9) - 6$; $3|-11| = -33$;

$33 = -33$ FALSE; Check $x = -\dfrac{5}{3}$: $3\left|2\left(-\dfrac{5}{3}\right) + 7\right| = 3\left(-\dfrac{5}{3}\right) - 6$;

$3\left|\dfrac{11}{3}\right| = -11$; $11 = -11$ FALSE; No solution.

1.16 $9x - 3 < -5x + 4$; $14x - 3 < 4$; $14x < 7$; $x < \dfrac{1}{2}$; The check is left to you.

1.17 $-2(x + 5) \geq \dfrac{1}{2}(5x - 2)$; $-4(x + 5) \geq 5x - 2$; $-4x - 20 \geq 5x - 2$;

$-20 \geq 9x - 2$; $-18 \geq 9x$; $-2 \geq x$ or $x \leq -2$; The check is left to you.

1.18 $\dfrac{7x+11}{3} \geq 6$ or $3(-2x + 1) > 15$; $\dfrac{7x+11}{3} \geq 6$; $7x + 11 \geq 18$;

$7x \geq 7$; $x \geq 1$; $3(-2x + 1) > 15$; $-6x + 3 > 15$; $-6x > 12$; $x < -2$; $x \geq 1$ or $x < -2$; The check is left to you.

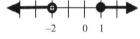

1.19 $3x + 1 > 5$ and $-2x + 6 > 4$; $3x + 1 > 5$; $3x > 4$; $x > \dfrac{4}{3}$

$-2x + 6 > 4$; $-2x > -2$; $x < 1$; $x > \dfrac{4}{3}$ and $x < 1$; No solution.

No numbers are greater than $1\dfrac{1}{3}$ and less than 1.

1.20 $3 < \dfrac{1}{3}(x+5) < 4$; $9 < x + 5 < 12$; $4 < x < 7$

The check is left to you.

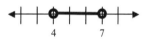

1.21 $|3x - 5| < 4$; $3x - 5 > -4$ and $3x - 5 < 4$; $3x - 5 > -4$; $3x > 1$;

$x > \dfrac{1}{3}$; $3x - 5 < 4$; $3x < 9$; $x < 3$; $x > \dfrac{1}{3}$ and $x < 3$

The check is left to you.

1.22 $-\dfrac{1}{2}|4x + 8| < 6$; $|4x + 8| > -12$; All distances are greater than or equal to 0, and are also greater than or equal to -12, so all real numbers are solutions.

Chapter 2

2.1

x	y
−1	−2
0	1
1	4

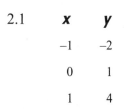

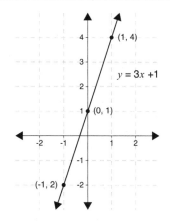

2.2 $x - y = 6$; $-y = -x + 6$; $y = x - 6$

x	y
−1	−7
0	−6
1	−5

2.3   slope $= 3$; y-intercept $= -1$

2.4 $m = \dfrac{3 - (-5)}{1 - 2} = \dfrac{8}{-1} = -8$

2.5   slope $= -\dfrac{3}{2}$, y-intercept $= 2$

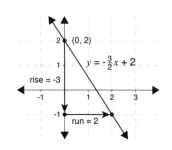

2.6 $x - 2y = 6;\ -2y = -x + 6;\ y = \dfrac{1}{2}x - 3;$

        slope $= \dfrac{1}{2}$, y-intercept $= -3$

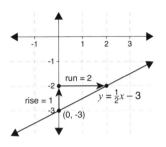

2.7 $y = -\dfrac{3}{2}x + 2$; $2y = -3x + 4$; $3x + 2y = 4$; $A = 3$, $B = 2$, $C = 4$

2.8 $y = 6$; $0x + y = 6$; $A = 0$, $B = 1$, $C = 6$

2.9 x-intercept: $y = 0$; $2x - 0 = -5$;

        $2x = -5$; $x = -\dfrac{5}{2}$; $\left(-\dfrac{5}{2}, 0\right)$

        y-intercept: $x = 0$;
        $2(0) - y = -5$;
        $-y = -5$; $y = 5$; $(0, 5)$

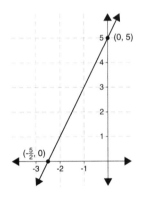

2.10 x-intercept: $y = 0$; $3x + 2(0) = 6$;
        $3x = 6$; $x = 2$; $(2, 0)$

        y-intercept: $x = 0$; $3(0) + 2y = 6$;
        $2y = 6$; $y = 3$; $(0, 3)$

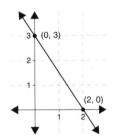

2.11   slope $= -3$, y-intercept $= 4$

2.12 $y = -3x + 4$ slope-intercept form; $3x + y = 4$ standard form

2.13 $y - 4 = 3(x - (-2))$; $y - 4 = 3x + 6$; $y = 3x + 10$; $-3x + y = 10$;
        $3x - y = -10$

2.14 $m = \dfrac{(y_2 - y_1)}{(x_2 - x_1)} = \dfrac{4 - (-3)}{1 - 2} = \dfrac{7}{-1} = -7$; $y - 4 = -7(x - 1)$;

$y - 4 = -7x + 7$; $7x + y = 11$

2.15   slope = 4; slope of perpendicular line = $-\dfrac{1}{4}$;

$y - (-3) = -\dfrac{1}{4}(x - 1)$; $y + 3 = -\dfrac{1}{4}(x - 1)$; $4y + 12 = -x + 1$;

$x + 4y = -11$

2.16 $2x + 3y = -6$; $3y = -2x - 6$; $y = -\dfrac{2}{3}x - 2$; slope $= -\dfrac{2}{3}$;

slope of parallel line $= -\dfrac{2}{3}$; $y - 4 = -\dfrac{2}{3}(x - (-1))$;

$3y - 12 = -2(x + 1)$; $3y - 12 = -2x - 2$; $2x + 3y = 10$

2.17 $\$80 = k(\$1,000)$; $0.08 = k$; Rate of interest 8%

$y = 0.08x$; $y = 0.08(\$4,500)$; $y = \$360$

2.18

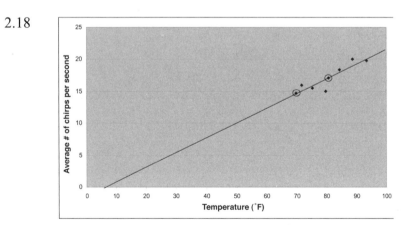

Answers will vary. Using the points (80.6, 17.1) and (69.7, 14.7):

$m = \dfrac{(y_2 - y_1)}{(x_2 - x_1)} = \dfrac{14.7 - 17.1}{69.7 - 80.6} = \dfrac{-2.4}{-10.9} \approx 0.22$

$y - 17.1 = 0.22(x - 80.6)$; $y - 17.1 = 0.22x - 17.732$;

$y = 0.22x - 0.632$

2.19 $y = 0.2327x - 1.64$

2.20 $y = 0.22x - 0.632$; $y = 0.22(65) - 0.632 = 13.668$
At 65 degrees, there will be approximately 14 chirps per second.

$y = 0.2327x - 1.64$; $y = 0.2327(65) - 1.64 = 13.4855$
At 65 degrees, there will be approximately 13 chirps per second.

2.21 $y = 0.22x - 0.632$; $22 = 0.22x - 0.632$; $22.632 = 0.22x$; $102.87 = x$
If there are 22 chirps per second, the estimated temperature is about 103°.

$y = 0.2327x - 1.64$; $22 = 0.2327x - 1.64$; $23.64 = 0.2327x$; $101.59 = x$
If there are 22 chirps per second, the estimated temperature is about 102°.

2.22 $y > -2x + 5$; $y = -2x + 5$;
slope $= -2$, y-intercept $= 5$

Check (0, 0): $0 > -2(0) + 5$;
$0 > 5$ FALSE

Check (2, 2): $2 > -2(2) + 5$;
$2 > 1$ ✓

(2, 2) lies in the region that is a solution, so shade the half-plane that contains (2, 2).

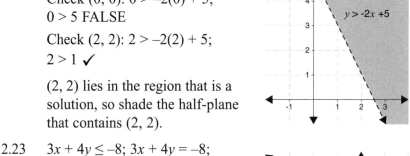

2.23 $3x + 4y \le -8$; $3x + 4y = -8$;

$4y = -3x - 8$; $y = -\dfrac{3}{4}x - 2$;

slope $= -\dfrac{3}{4}$, y-intercept $= -2$

Check (0, 0): $3(0) + 4(0) \le -8$;
$0 \le -8$ FALSE

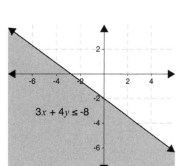

Check (−2, −2): $3(-2) + 4(-2) \le -8$; $-14 \le -8$ ✓; (−2, −2) lies in the region that is a solution, so shade the half-plane that contains (−2, −2).

Chapter 3

3.1 $x + 2y = -4$; $2y = -x - 4$;

$y = -\dfrac{1}{2}x - 2$; slope = $-\dfrac{1}{2}$,

y-intercept = -2

$3x - y = 2$; $-y = -3x + 2$;
$y = 3x - 2$; slope = 3,
y-intercept = -2

Solution: $(0, -2)$, consistent
and independent

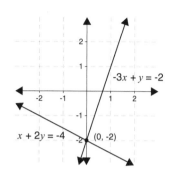

3.2 $2x - y = -4$; $-y = -2x - 4$; $y = 2x + 4$;
slope = 2, y-intercept = 4

$-6x + 3y = 12$; $3y = 6x + 12$; $y = 2x + 4$;
slope = 2, y-intercept = 4

The lines are the same and the system
has an infinite number of solutions. The
system is consistent and dependent.

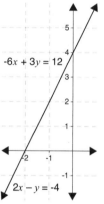

3.3 $3x + y = 7$; $y = -3x + 7$; $2x + 3(-3x + 7) = 7$; $2x - 9x + 21 = 7$;
$-7x = -14$; $x = 2$;

$3(2) + y = 7$; $6 + y = 7$; $y = 1$; $(2, 1)$

3.4 $3x + y = 2$; $y = -3x + 2$; $4x + 2(-3x + 2) = 4$; $4x - 6x + 4 = 4$;
$-2x = 0$; $x = 0$;

$3(0) + y = 2$; $y = 2$; $(0, 2)$

3.5 $5x - 2y = 27 \quad \rightarrow \qquad 10x - 4y = 54$
$$\underline{-3x + 4y = -5}$$
$$7x = 49;\; x = 7$$

$5(7) - 2y = 27$; $35 - 2y = 27$; $-2y = -8$; $y = 4$; $(7, 4)$

3.6 $2x + 6y = 26 \quad \rightarrow \qquad 4x + 12y = \quad 52$
$7x - 4y = -34 \rightarrow \quad \underline{21x - 12y = -102}$
$$25x = \quad -50;\; x = -2$$

$2(-2) + 6y = 26$; $-4 + 6y = 26$; $6y = 30$; $y = 5$; $(-2, 5)$

3.7 $2x - y = 6$ → $4x - 2y = 12$

        $\underline{-4x + 2y = \;\; 3}$

                    $0 = 15$ FALSE

There is no solution. The system is inconsistent.

3.8 $4x - 3y = -6$ → $-20x + 15y = \;\; 30$

        $5x - 2y = -11$ → $\underline{20x \; - \; 8y = -44}$

                                $7y = -14; y = -2$

$4x - 3(-2) = -6; 4x + 6 = -6; 4x = -12; x = -3; (-3, -2)$.
The system is consistent and independent.

3.9 $x + 3y = 10; x = -3y + 10; -2x - 6y = -20$;
        $-2(-3y + 10) - 6y = -20; 6y - 20 - 6y = -20; -20 = -20$

TRUE; There are an infinite number of solutions. The system
is consistent and dependent.

3.10 $y = \dfrac{1}{2}x + 1$: slope $= \dfrac{1}{2}$, y-intercept $= 1$;

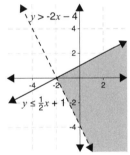

Check $(0, 0)$: $y \le \dfrac{1}{2}x + 1$; $0 \le \dfrac{1}{2}(0) + 1$;

$0 \le 0 + 1; 0 \le 1$ ✓

Shade the region that includes $(0, 0)$.

$y = -2x - 4$: slope $= -2$, y-intercept $= -4$
Check $(0, 0)$: $y > -2x - 4$; $0 > -2(0) - 4$; $0 > -4$ ✓
Shade the region that includes $(0, 0)$.

3.11 $y = 3x + 1$: slope $= 3$,
        y-intercept $= 1$; Check $(0, 0)$:
        $y < 3x + 1$; $0 < 3(0) + 1$; $0 < 1$ ✓
        Shade the region that includes
        $(0, 0)$. $-3x + y = -4$: $y = 3x - 4$;
        slope $= 3$, y-intercept $= -4$

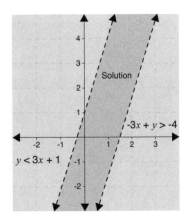

Check $(0, 0)$: $-3x + y > -4$;
$-3(0) + 0 > -4$; $0 > -4$ ✓

Shade the region that includes
$(0, 0)$.

3.12 $-2x + y = -3$; $y = 2x - 3$; slope = 2, y-intercept = -3

Check $(0, 0)$: $-2x + y > -3$; $-2(0) + 0 > -3$; $0 > -3$ TRUE

Shade the region that includes $(0, 0)$. $y = 5$; slope = 0,

y-intercept = 5; Check $(0, 0)$: $y < 5$; $0 < 5$ ✓

Shade the region that includes $(0, 0)$. $4x + 3y = 3$;

$3y = -4x + 3$; $y = -\dfrac{4}{3}x + 1$; slope = $-\dfrac{4}{3}$, y-intercept = 1

Check $(0, 0)$: $4x + 3y > 3$; $4(0) + 3(0) > 3$; $0 > 3$ FALSE
Shade the region that does not include $(0, 0)$.

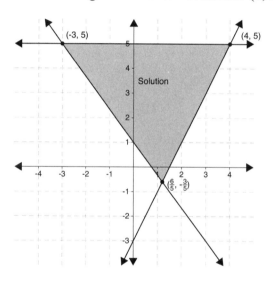

Intersection of $4x + 3y = 3$ and $-2x + y = -3$:

$$-2x + y = -3 \rightarrow \qquad -4x + 2y = -6$$
$$\underline{4x + 3y = 3}$$
$$5y = -3; \; y = -\dfrac{3}{5}$$

$-2x - \dfrac{3}{5} = -3$; $-2x = -\dfrac{12}{5}$; $x = \dfrac{6}{5}$; Intersection points $(-3, 5)$,

$(4, 5)$, $\left(\dfrac{6}{5}, -\dfrac{3}{5} \right)$.

3.13 $x + 2y = 10$; $2y = -x + 10$; $y = -\dfrac{1}{2}x + 5$; slope $= -\dfrac{1}{2}$,

y-intercept $= 5$; $x - 4y = -8$; $-4y = -x - 8$; $y = \dfrac{1}{4}x + 2$; slope $= \dfrac{1}{4}$,

y-intercept $= 2$; $y = 3x - 2$; slope $= 3$, y-intercept $= -2$

Find intersection of $y = 3x - 2$ and
$x - 4y = -8$. $x - 4(3x - 2) = -8$;
$x - 12x + 8 = -8$; $-11x = -16$;

$x = \dfrac{16}{11} \approx 1.45$; $y = 3(1.45) - 2 = 2.35$;

Coordinates of vertices $= (2, 4)$, $(4, 3)$,
$(1.45, 2.35)$; $P = 4x + 18y$; $P = 4(2) +$
$18(4) = 80$; $P = 4(4) + 18(3) = 70$;
$P = 4(1.45) + 18(2.35) = 48.1$;

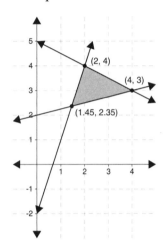

Minimum value is 48.1 given by
coordinates $(1.45, 2.35)$, and the
maximum value is 80 given by the
coordinates $(2, 4)$.

3.14 $2x + 4y = 40$; $4y = -2x + 40$; $y = -\dfrac{1}{2} + 10$

slope $= -\dfrac{1}{2}$, y-intercept $= 10$

$y = 2x$; slope $= 2$, y-intercept $= 0$
Coordinates of vertices $= (0, 10)$,
$(4, 8)$, $(0, 0)$; $P = 25x + 40y$
$P = 25(0) + 40(10) = 400$
$P = 25(4) + 40(8) = 420$
$P = 25(0) + 40(0) = 0$

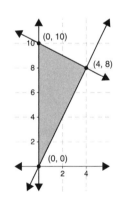

The maximum profit of $420 is earned if she makes and sells
four of the first design and eight of the second design.

3.15   To eliminate y, multiply Equation 1 by 2 and add to Equation 2.

$2x - y + 5z = 25 \qquad \rightarrow \qquad 4x - 2y + 10z = 50$
$$\underline{-3x + 2y + z = -4}$$
$$x + 11z = 46$$

Multiply Equation 1 by -3 and add to Equation 3.

$2x - y + 5z = 25 \qquad \rightarrow \qquad -6x + 3y - 15z = -75$
$$\underline{6x - 3y + 2z = \ \ 23}$$
$$-13z = -52; \, z = 4$$

$x + 11(4) = 46; \, x + 44 = 46; \, x = 2; \, 2(2) - y + 5(4) = 25;$
$4 - y + 20 = 25; \, 24 - y = 25; \, -y = 1; \, y = -1; \, (2, -1, 4);$
Check: $2(2) - (-1) + 5(4) = 25; \, 25 = 25 \checkmark;$
$-3(2) + 2(-1) + 4 = -4; \, -4 = -4 \checkmark; \, 6(2) - 3(-1) + 2(4) = 23;$
$23 = 23 \checkmark;$ The solution to the system is $(2, -1, 4)$.

3.16   To eliminate x, add Equation 2 and Equation 3.

$-x + 3y - 2z = -1$
$\underline{x - 2y + 3z = \ \ 3}$
$\qquad y + z = \ \ 2$

To eliminate x, multiply Equation 3 by -2 and add to Equation 1.

$x - 2y + 3z = 3 \qquad \rightarrow \qquad -2x + 4y - 6z = -6$
$$\underline{2x - 4y + 6z = \ \ 5}$$
$$0 = -1 \text{ FALSE}$$

The system has no solution.

3.17   To eliminate z, add Equation 1 and Equation 3.

$x + y + z = 0$
$\underline{x + y - z = 5}$
$2x + 2y = 5$

To eliminate z, multiply Equation 3 by 2 and add to Equation 2.

$x + y - z = 5 \qquad \rightarrow \qquad 2x + 2y - 2z = 10$
$$\underline{4x + 4y + 2z = \ \ 5}$$
$$6x + 6y = 15$$

Solve the system $\begin{cases} 2x + 2y = 5 & \text{Equation 4} \\ 6x + 6y = 15 & \text{Equation 5} \end{cases}$

Multiply Equation 4 by –3 and add to Equation 5.

$$2x + 2y = 5 \quad \rightarrow \quad \begin{array}{r} -6x - 6y = -15 \\ 6x + 6y = 15 \\ \hline 0 = 0 \end{array}$$

Equations 1 and 3 are not the same plane, and are not parallel, because the variables didn't cancel out in the first pair of equations. Equations 2 and 3 are not the same plane, and are not parallel, because the variables didn't cancel out in the second pair of equations. The system has infinitely many solutions.

3.18
$$\left[\begin{array}{ccc|c} 3 & 1 & 1 & 7 \\ 2 & -3 & -2 & -6 \\ 1 & 4 & 1 & 5 \end{array}\right]; \left[\begin{array}{ccc|c} 1 & 4 & 1 & 5 \\ 2 & -3 & -2 & -6 \\ 3 & 1 & 1 & 7 \end{array}\right]$$ Exchange Row 1 and Row 3;

$$\left[\begin{array}{ccc|c} 1 & 4 & 1 & 5 \\ 0 & -11 & -4 & -16 \\ 3 & 1 & 1 & 7 \end{array}\right]$$ New Row 2 = –2(Row 1) + Row 2

$$\left[\begin{array}{ccc|c} 1 & 4 & 1 & 5 \\ 0 & -11 & -4 & -16 \\ 0 & -11 & -2 & -8 \end{array}\right]$$ New Row 3 = –3(Row 1) + Row 3

$$\left[\begin{array}{ccc|c} 1 & 4 & 1 & 5 \\ 0 & -11 & -4 & -16 \\ 0 & 0 & 2 & 8 \end{array}\right]$$ New Row 3 = –1(Row 2) + Row 3

$$\left[\begin{array}{ccc|c} 1 & 4 & 1 & 5 \\ 0 & 1 & \frac{4}{11} & \frac{16}{11} \\ 0 & 0 & 1 & 4 \end{array}\right]$$ New Row 2 = $-\frac{1}{11}$(Row 2), New Row 3 = $\frac{1}{2}$

$$z = 4; \quad y + \frac{4}{11}z = \frac{16}{11}; \quad y + \frac{4}{11}(4) = \frac{16}{11}; \quad y + \frac{16}{11} = \frac{16}{11}; \quad y = 0$$

$x + 4y + z = 5; x + 4(0) + 4 = 5; x + 4 = 5; x = 1$

The solution to the system is $(1, 0, 4)$. The check is left to you.

3.19
$$\begin{bmatrix} 2 & -1 & 5 & | & 25 \\ -3 & 2 & 1 & | & -4 \\ 6 & -3 & 2 & | & 23 \end{bmatrix}; \begin{bmatrix} -1 & 1 & 6 & | & 21 \\ -3 & 2 & 1 & | & -4 \\ 6 & -3 & 2 & | & 23 \end{bmatrix}$$ New Row 1 = Row 1 + Row 2

$$\begin{bmatrix} -1 & 1 & 6 & | & 21 \\ -3 & 2 & 1 & | & -4 \\ 0 & 3 & 38 & | & 149 \end{bmatrix}$$ New Row 3 = 6(Row 1) + Row 3

$$\begin{bmatrix} -1 & 1 & 6 & | & 21 \\ 0 & -1 & -17 & | & -67 \\ 0 & 3 & 38 & | & 149 \end{bmatrix}$$ New Row 2 = −3(Row 1) + Row 2

$$\begin{bmatrix} -1 & 1 & 6 & | & 21 \\ 0 & -1 & -17 & | & -67 \\ 0 & 0 & -13 & | & -52 \end{bmatrix}$$ New Row 3 = 3(Row 2) + Row 3

$$\begin{bmatrix} 1 & -1 & -6 & | & -21 \\ 0 & 1 & 17 & | & 67 \\ 0 & 0 & 1 & | & 4 \end{bmatrix}$$ New Row 1 = −1(Row 1),
New Row 2 = −1(Row 2)
New Row 3 = $-\dfrac{1}{13}$ (Row 3)

$z = 4; y + 17z = 67; y + 17(4) = 67; y + 68 = 67; y = -1;$
$x - y - 6z = -21; x - (-1) - 6(4) = -21; x - 23 = -21; x = 2;$ The solution is $(2, -1, 4)$. See the solution and check for Exercise 3.15.

3.20 $D = \begin{vmatrix} 2 & 3 \\ 1 & -4 \end{vmatrix} = 2(-4) - 1(3) = -11$ $D \neq 0$, so there is a unique solution.

$D_x = \begin{vmatrix} -1 & 3 \\ 16 & -4 \end{vmatrix} = -1(-4) - 16(3) = -44$;

$D_y = \begin{vmatrix} 2 & -1 \\ 1 & 16 \end{vmatrix} = 2(16) - 1(-1) = 33$

$$x = \frac{D_x}{D} = \frac{-44}{-11} = 4; \; y = \frac{D_y}{D} = \frac{33}{-11} = -3; \; (4, -3).$$

The check is left to you.

3.21 $D = \begin{vmatrix} 2 & -1 & 5 \\ -3 & 2 & 1 \\ 6 & -3 & 2 \end{vmatrix} = 2\begin{vmatrix} 2 & 1 \\ -3 & 2 \end{vmatrix} - (-1)\begin{vmatrix} -3 & 1 \\ 6 & 2 \end{vmatrix} + 5\begin{vmatrix} -3 & 2 \\ 6 & -3 \end{vmatrix} = -13$

$D_x = \begin{vmatrix} 25 & -1 & 5 \\ -4 & 2 & 1 \\ 23 & -3 & 2 \end{vmatrix} = 25\begin{vmatrix} 2 & 1 \\ -3 & 2 \end{vmatrix} - (-1)\begin{vmatrix} -4 & 1 \\ 23 & 2 \end{vmatrix} + 5\begin{vmatrix} -4 & 2 \\ 23 & -3 \end{vmatrix} = -26$

$D_y = \begin{vmatrix} 2 & 25 & 5 \\ -3 & -4 & 1 \\ 6 & 23 & 2 \end{vmatrix} = 2\begin{vmatrix} -4 & 1 \\ 23 & 2 \end{vmatrix} - 25\begin{vmatrix} -3 & 1 \\ 6 & 2 \end{vmatrix} + 5\begin{vmatrix} -3 & -4 \\ 6 & 23 \end{vmatrix} = 13$

$D_z = \begin{vmatrix} 2 & -1 & 25 \\ -3 & 2 & -4 \\ 6 & -3 & 23 \end{vmatrix} = 2\begin{vmatrix} 2 & -4 \\ -3 & 23 \end{vmatrix} - (-1)\begin{vmatrix} -3 & -4 \\ 6 & 23 \end{vmatrix} + 25\begin{vmatrix} -3 & 2 \\ 6 & -3 \end{vmatrix} = -52$

$$x = \frac{D_x}{D} = \frac{-26}{-13} = 2; \; y = \frac{D_y}{D} = \frac{13}{-13} = -1; \; z = \frac{D_z}{D} = \frac{-52}{-13} = 4$$

The solution to the system is (2, –1, 4). See the solution and check for Exercises 3.15 and 3.19.

Chapter 4

4.1   Domain = {–3, 2, 4, 5}, Range = {0, 1, 2}

4.2   Domain = {all real numbers ≥ 0}, Range = {all real numbers}

4.3   Yes, the relation is a function because none of the domain values are repeated.

4.4   No, the relation is not a function because the domain value 1 is paired with two different range values: (1, 3) and (1, 0).

4.5   Yes, the relation is a function. Any non-vertical line is a function.

4.6

y	y²	x	(x, y)
0	0	0	(0, 0)
1	1	1	(1, 1)
−1	1	1	(1, −1)
2	4	4	(4, 2)
−2	4	4	(4, −2)

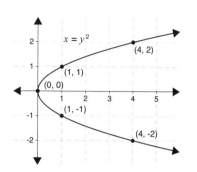

The relation is not a function.
It fails the vertical line test.

4.7 $f(0) = 0^2 + 2(0) - 3 = -3$

4.8 $f(x - h) = -(x - h) + 3 = -x + h + 3$

4.9 $f(4y) = 6(4y) + 2 = 24y + 2$

4.10 $f(x) = \dfrac{2}{5}x + 3$ is equivalent to

$y = \dfrac{2}{5}x + 3$; slope $= m = \dfrac{2}{5}$;

y-intercept $= b = 3$.

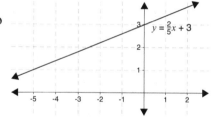

4.11

x	−x² − 4	f(x)
−2	$-(-2)^2 - 4$	−8
−1	$-(-1)^2 - 4$	−5
0	$-(0)^2 - 4$	−4
1	$-(1)^2 - 4$	−5
2	$-(2)^2 - 4$	−8

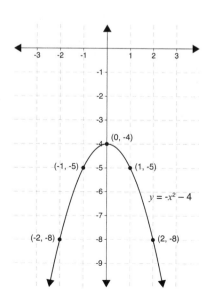

4.12 $f(x) = x - 3$ if $x \le -1$ $f(x) = x + 2$ if $x > -1$

x	x − 3	f(x)
−1	−1 − 3	−4
−2	−2 − 3	−5
−3	−3 − 3	−6

x	x + 2	f(x)
−1	−1 + 2	1
0	0 + 2	2
1	1 + 2	3

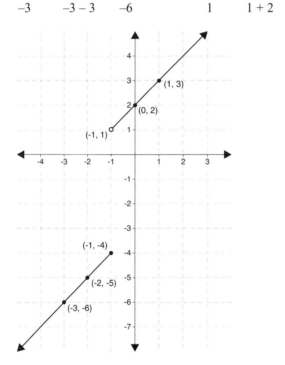

The domain is all real numbers. The range is all real numbers ≤ -4 and > 1.

4.13

x	|x − 2|	f(x)
0	|0 − 2|	2
1	|1 − 2|	1
2	|2 − 2|	0
3	|3 − 2|	1
4	|4 − 2|	2

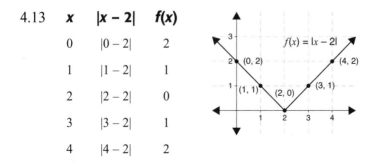

The domain is all real numbers. The range is $f(x) \ge 0$.

4.14

Hours	Charge
$0 < x \le 1$	$45
$1 < x \le 2$	$90
$2 < x \le 3$	$135
$3 < x \le 4$	$180
$4 < x \le 5$	$225

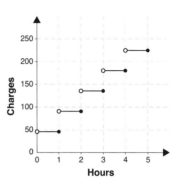

4.15 $y = |x| - 1$ is of the form
$y = f(x) + k$ where $k = -1$.
The parent function $f(x) = |x|$
will be translated 1 unit down.

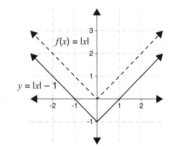

4.16 $y = (x + 4)^2$ is of the form
$y = f(x - h)$, where $h = -4$.
The parent function
$f(x) = x^2$ will be translated
4 units to the left.

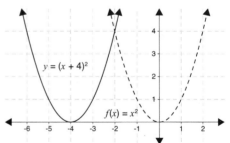

4.17 $y = -(x + 2)^2$ is of the form
$y = -f(x - h)$. The parent function
$f(x) = x^2$ will be translated 2 units
to the left and reflected over the
x-axis.

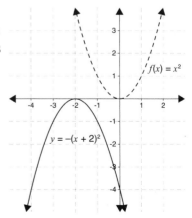

4.18 $y = -|x + 3|$ is of the form
$y = -f(x - h)$. The parent function
$f(x) = |x|$ will be translated 3 units
to the left and reflected over the
x-axis.

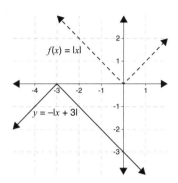

4.19 $y = -3x^2$ is of the form $y = af(x)$,
where $a < 0$ and $|a| > 1$. The parent
function $f(x) = x^2$ will be reflected
over the x-axis and stretched
vertically.

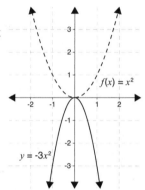

4.20 $y = \frac{1}{3}|x|$ is of the form $y = af(x)$,

where $a > 0$ and $0 < |a| < 1$. The
parent function $f(x) = |x|$ will be
compressed vertically.

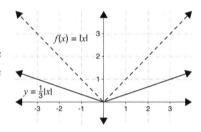

4.21   The function is an
absolute function. The
parent function $f(x) = |x|$
is translated 1 unit to the
left. It is reflected over
the x-axis and translated
3 units down.

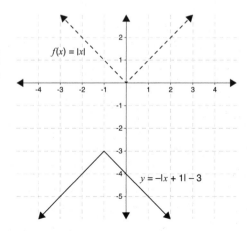

4.22 $f(x) + g(x) = 2x + 3 + x = 3x + 3$; $f(x) - g(x) = 2x + 3 - x = x + 3$

$(f \cdot g)(x) = (2x + 3)(x) = 2x^2 + 3x$; $\dfrac{f}{g}(x) = \dfrac{2x+3}{x}$, $x \neq 0$

4.23 $f(x) + g(x) = x + 2x + 3 = 3x + 3$; $f(x) - g(x) = x - (2x + 3) = -x - 3$

$(f \cdot g)(x) = (x)(2x + 3) = 2x^2 + 3x$; $\dfrac{f}{g}(x) = \dfrac{x}{2x+3}$, $x \neq -\dfrac{3}{2}$

4.24 $(f \circ g)(x) = f(g(x)) = f(2x - 3) = \dfrac{1}{2x-3}$; $(f \circ g)(2) = \dfrac{1}{2(2)-3} = 1$

Restrictions on the domain of g: none; Restrictions on the domain of f: $x \neq 0$. So $g(x) \neq 0$; Solve $g(x) = 0$ to find the restricted value.

$2x - 3 = 0$; $x = \dfrac{3}{2}$; Domain of the composite function:

all real numbers except $\dfrac{3}{2}$.

4.25 $(g \circ f)(x) = g(f(x)) = g\left(\dfrac{1}{x}\right) = 2\left(\dfrac{1}{x}\right) - 3 = \dfrac{2}{x} - 3$

$(g \circ f)(-2) = \dfrac{2}{-2} - 3 = -4$; Restrictions on the domain of f: $x \neq 0$;

Restrictions on the domain of g: none; Domain of the composite function: all real numbers except 0.

4.26 $(f \circ g)(x) = f(g(x)) = 3(x + 3) - 1 = 3x + 9 - 1 = 3x + 8$

$(g \circ f)(x) = g(f(x)) = 3x - 1 + 3 = 3x + 2$; Since $(f \circ g)(x) \neq$

$(g \circ f)(x) \neq x$, $f(x)$ and $g(x)$ are not inverses.

4.27 $y = 6x + 2$; $x = 6y + 2$; $x - 2 = 6y$; $y = \dfrac{x-2}{6}$

Chapter 5

5.1 $15x^2 - 6x = 0$; $3x(5x - 2) = 0$; $3x = 0$ or $5x - 2 = 0$; $x = 0$ or $x = \dfrac{2}{5}$

5.2 $16x^2 - 36 = 0$; $(4x + 6)(4x - 6) = 0$; $4x + 6 = 0$ or $4x - 6 = 0$;

$x = -\dfrac{3}{2}$ or $x = \dfrac{3}{2}$

5.3 $4x^2 + 4x + 1 = 0$; $(2x + 1)^2 = 0$; $2x + 1 = 0$; $x = -\dfrac{1}{2}$

5.4 $2x^2 - 5x - 12 = 0$; $(2x + 3)(x - 4) = 0$; $2x + 3 = 0$ or $x - 4 = 0$;

$x = -\dfrac{3}{2}$ or $x = 4$

5.5 $15x^2 + 39x - 18 = 0$; $3(5x^2 + 13x - 6) = 0$; $3(5x - 2)(x + 3) = 0$;

$5x - 2 = 0$ or $x + 3 = 0$; $x = \dfrac{2}{5}$ or $x = -3$

5.6 $x^2 + 10x + 20 = 0$; $x^2 + 10x = -20$;

$x^2 + 10x + 25 = -20 + 25$; $(x + 5)^2 = 5$; $x + 5 = \pm\sqrt{5}$; $x = -5 \pm \sqrt{5}$

Check $x = -5 + \sqrt{5}$: $(-5 + \sqrt{5})^2 + 10(-5 + \sqrt{5}) + 20 = 0$;

$25 - 10\sqrt{5} + 5 - 50 + 10\sqrt{5} + 20 = 0$; $0 = 0$ ✓

Check $x = -5 + \sqrt{5}$: $(-5 - \sqrt{5})^2 + 10(-5 - \sqrt{5}) + 20 = 0$;

$25 + 10\sqrt{5} + 5 - 50 - 10\sqrt{5} + 20 = 0$; $0 = 0$ ✓

5.7 $3x^2 - 9x - 27 = 0$; $x^2 - 3x - 9 = 0$; $x^2 - 3x = 9$;

$x^2 - 3x + \dfrac{9}{4} = 9 + \dfrac{9}{4}$; $\left(x - \dfrac{3}{2}\right)^2 = \dfrac{45}{4}$; $x - \dfrac{3}{2} = \pm\sqrt{\dfrac{45}{4}}$;

$x = \dfrac{3}{2} \pm \sqrt{\dfrac{45}{4}} = \dfrac{3}{2} \pm \dfrac{\sqrt{9} \cdot \sqrt{5}}{2} = \dfrac{3 \pm 3\sqrt{5}}{2}$. The check is left to you.

5.8 $\sqrt{-243} = \sqrt{-1} \cdot \sqrt{81} \cdot \sqrt{3} = 9i\sqrt{3}$

5.9 $i^{25} = i^8 \cdot i^8 \cdot i^8 \cdot i = 1 \cdot 1 \cdot 1 \cdot i = i$

5.10 $2i \cdot (-7i) = -14i^2 = -14(-1) = 14$

5.11 $\sqrt{-12} \cdot \sqrt{-3} = i\sqrt{12} \cdot i\sqrt{3} = i^2 \sqrt{36} = -6$

5.12 $x^2 + 20 = 0$; $x^2 = -20$; $x = \pm\sqrt{-20} = \pm i\sqrt{4} \cdot \sqrt{5} = \pm 2i\sqrt{5}$

Check $x = 2i\sqrt{5}$: $(2i\sqrt{5})^2 + 20 = 0$; $4i^2(5) + 20 = 0$; $-20 + 20 = 0$ ✓

Check $x = -2i\sqrt{5}$: $(-2i\sqrt{5})^2 + 20 = 0$; $4i^2(5) + 20 = 0$;
$-20 + 20 = 0$ ✓

5.13 $x^2 - 8x + 25 = 0$; $x^2 - 8x = -25$; $x^2 - 8x + 16 = -25 + 16$;
$(x-4)^2 = -9$; $x - 4 = \pm\sqrt{-9}$; $x = 4 \pm \sqrt{-9} = 4 \pm i\sqrt{9} = 4 \pm 3i$
The check is left to you.

5.14 $x^2 + 5x + 7 = 0$; $x^2 + 5x = -7$; $x^2 + 5x + \dfrac{25}{4} = -7 + \dfrac{25}{4}$;

$\left(x + \dfrac{5}{2}\right)^2 = -\dfrac{3}{4}$; $x + \dfrac{5}{2} = \pm\sqrt{-\dfrac{3}{4}}$;

$x = -\dfrac{5}{2} \pm \sqrt{-\dfrac{3}{4}} = -\dfrac{5}{2} \pm \dfrac{i\sqrt{3}}{2} = \dfrac{-5 \pm i\sqrt{3}}{2}$. The check is left to you.

5.15 $x^2 + 5x - 3 = 0$; $a = 1$, $b = 5$, $c = -3$; $x = \dfrac{-(5) \pm \sqrt{(5)^2 - 4(1)(-3)}}{2(1)}$;

$x = \dfrac{-5 \pm \sqrt{25 + 12}}{2}$; $x = \dfrac{-5 \pm \sqrt{37}}{2}$

5.16 $2x^2 - 7x + 14 = 0$; $a = 2$, $b = -7$, $c = 14$; $x = \dfrac{-(-7) \pm \sqrt{(-7)^2 - 4(2)(14)}}{2(2)}$;

$x = \dfrac{7 \pm \sqrt{49 - 112}}{4}$; $x = \dfrac{7 \pm \sqrt{-63}}{4} = \dfrac{7 \pm \sqrt{-1} \cdot \sqrt{9} \cdot \sqrt{7}}{4} = \dfrac{7 \pm 3i\sqrt{7}}{4}$

5.17 $x^2 - 6x + 8 = 0$; $(x - 4)(x - 2) = 0$; $x - 4 = 0$ or $x - 2 = 0$; $x = 4$
or $x = 2$

5.18 $x^2 + 10x - 3 = 0$; $x^2 + 10x = 3$; $x^2 + 10x + 25 = 3 + 25$;
$(x + 5)^2 = 28$; $x + 5 = \pm\sqrt{28}$; $x = -5 \pm \sqrt{4} \cdot \sqrt{7}$; $x = -5 \pm 2\sqrt{7}$

5.19 $x^2 - 5x + 9 = 0$; $a = 1$, $b = -5$, $c = 9$; $x = \dfrac{-(-5) \pm \sqrt{(-5)^2 - 4(1)(9)}}{2(1)}$;

$x = \dfrac{5 \pm \sqrt{25 - 36}}{2}$; $x = \dfrac{5 \pm \sqrt{-11}}{2}$; $x = \dfrac{5 \pm i\sqrt{11}}{2}$

5.20 $x^2 + 4x - 10 = 0$: $a = 1$, $b = 4$, $c = -10$

$b^2 - 4ac = 4^2 - 4(1)(-10) = 16 + 40 = 56$

There are two real solutions.

5.21 $3x^2 - 5x + 6 = 0$: $a = 3$, $b = -5$, $c = 6$

$b^2 - 4ac = (-5)^2 - 4(3)(6) = 25 - 72 = -47$

There are two complex solutions.

5.22 $2x^2 - 7x - 4 \le 0$; $2x^2 - 7x - 4 = 0$; $(2x + 1)(x - 4) = 0$; $2x + 1 = 0$

or $x - 4 = 0$; $x = -\dfrac{1}{2}$ or $x = 4$. Choose points from each region.

$x = -1$: $2(-1)^2 - 7(-1) - 4 \le 0$; $5 \le 0$; $x = 0$; $2(0)^2 - 7(0) - 4 \le 0$;

$-4 \le 0$ ✓ $x = 5$: $2(5)^2 - 7(5) - 4 \le 0$; $11 \le 0$. The solution to the

quadratic inequality is $-\dfrac{1}{2} \le x \le 4$.

In set form, $\left\{ x \middle| -\dfrac{1}{2} \le x \le 4 \right\}$. In interval form, $\left[-\dfrac{1}{2}, 4 \right]$.

5.23 $x^2 + 5x > 1$; $x^2 + 5x = 1$; $x^2 + 5x - 1 = 0$; $a = 1$, $b = 5$, $c = -1$

$$x = \frac{-5 \pm \sqrt{(5)^2 - 4(1)(-1)}}{2(1)}; \ x = \frac{-5 \pm \sqrt{25 + 4}}{2}; \ x = \frac{-5 \pm \sqrt{29}}{2}$$

$x \approx -5.1926$ or $x \approx 0.1926$. Choose points from each region.

$x = -6$: $(-6)^2 + 5(-6) > 1$; $6 > 1$ ✓ $x = 0$; $(0)^2 + 5(0) > 1$; $0 > 1$

$x = 1$: $(1)^2 + 5(1) > 1$; $6 > 1$ ✓ The solution to the quadratic

inequality is $x < \dfrac{-5 - \sqrt{29}}{2}$ or $x > \dfrac{-5 + \sqrt{29}}{2}$.

In set form, $\left\{ x \middle| x < \dfrac{-5 - \sqrt{29}}{2} \text{ or } x > \dfrac{-5 + \sqrt{29}}{2} \right\}$.

In interval form, $\left(-\infty, \dfrac{-5 - \sqrt{29}}{2} \right) \cup \left(\dfrac{-5 + \sqrt{29}}{2}, \infty \right)$.

5.24

x	$x^2 - 6x + 10$	f(x)	(x, f(x))
0	$(0)^2 - 6(0) + 10$	10	(0, 10)
1	$(1)^2 - 6(1) + 10$	5	(1, 5)
3	$(3)^2 - 6(3) + 10$	1	(3, 1)
5	$(5)^2 - 6(5) + 10$	5	(5, 5)
6	$(6)^2 - 6(6) + 10$	10	(6, 10)

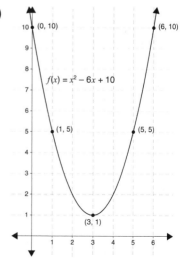

5.25

x	$-2x^2 - 4x + 5$	$f(x)$	$(x, f(x))$
-3	$-2(-3)^2 - 4(-3) + 5$	-1	$(-3, -1)$
-2	$-2(-2)^2 - 4(-2) + 5$	5	$(-2, 5)$
-1	$-2(-1)^2 - 4(-1) + 5$	7	$(-1, 7)$
0	$-2(0)^2 - 4(0) + 5$	5	$(0, 5)$
1	$-2(1)^2 - 4(1) + 5$	-1	$(1, -1)$

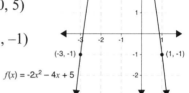

5.26 Line of symmetry:

$$x = \frac{-b}{2a} = \frac{-(4)}{2(1)} = -2$$

x-coordinate of vertex:

$$\frac{-b}{2a} = \frac{-(4)}{2(1)} = -2$$

y-coordinate of vertex:

$$f(x) = (-2)^2 + 4(-2) + 1 = -3$$

vertex $= (-2, -3)$; y-intercept $= c = 1$

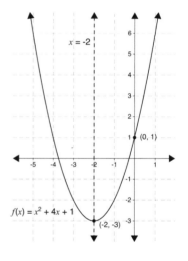

5.27 Line of symmetry: $x = \dfrac{-b}{2a} = \dfrac{-(-8)}{2(2)} = 2$

x-coordinate of vertex: $\dfrac{-b}{2a} = \dfrac{-(-8)}{2(2)} = 2$

y-coordinate of vertex:

$$f(x) = 2(2)^2 - 8(2) + 5 = -3$$

vertex $= (2, -3)$; y-intercept $= c = 5$

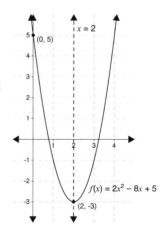

5.28 $h = 1$, $k = -3$, so the vertex is $(1, -3)$. The line of symmetry is $x = 1$. $|a| > 1$, so the parabola is stretched vertically.

When $x = 0$, $y = 2(0-1)^2 - 3 = -1$, so $(0, -1)$ is on the parabola. When $x = 2$, $y = 2(2-1)^2 - 3 = -1$, so $(2, -1)$ is on the parabola.

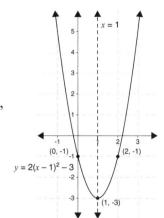

5.29 $y = x^2 - 4x + 9$; $y = (x^2 - 4x + 4) + 5$;

$y = (x-2)^2 + 5$; $h = 2$ and $k = 5$, so the vertex of the parabola is $(2, 5)$. The line of symmetry is $x = 2$. $a = 1$, so the shape and direction of the parabola is the same as the parent function $f(x) = x^2$.

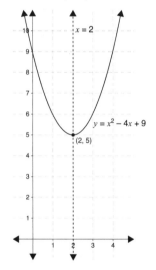

5.30 $y = 3x^2 + 18x - 5$; $y = 3(x^2 + 6x) - 5$;

$y = 3(x^2 + 6x) + 3(9) - 3(9) - 5$;

$y = 3(x^2 + 6x + 9) - 32$; $y = 3(x+3)^2 - 32$;

$h = -3$ and $k = -32$, so the vertex of the parabola is $(-3, -32)$. The line of symmetry is $x = -3$. $a = 3$, so the direction of the parabola is the same as the parent function $f(x) = x^2$, but the parabola is stretched vertically.

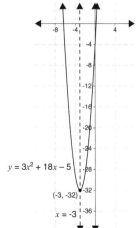

Chapter 6

6.1 $2n^{-3}\left(\dfrac{1}{4}n^4\right) = \dfrac{1n^4}{2n^3} = \dfrac{n^{4-3}}{2} = \dfrac{n}{2}$

6.2 $\left(6a^{-3}b^0\right)^2 = 6^2\left(a^{-3}\right)^2\left(b^0\right)^2 = 36a^{-6}b^0 = \dfrac{36}{a^6}$

6.3 $\dfrac{5^{-1}y^2}{5y^6} = \dfrac{y^2}{25y^6} = \dfrac{y^{2-6}}{25} = \dfrac{y^{-4}}{25} = \dfrac{1}{25y^4}$

6.4 $\dfrac{\left(2x^5y^{-2}\right)^3\left(z^3\right)^0}{8x^2y^{-1}z^2} = \dfrac{2^3\left(x^5\right)^3\left(y^{-2}\right)^3\left(z^3\right)^0}{8x^2y^{-1}z^2} = \dfrac{8x^{15}y^{-6}z^0}{8x^2y^{-1}z^2} = \dfrac{x^{15}y^1}{x^2y^6z^2}$

       $= \dfrac{x^{15-2}y^{1-6}}{z^2} = \dfrac{x^{13}y^{-5}}{z^2} = \dfrac{x^{13}}{y^5z^2}$

6.5 $(4x^2y - 3xy - 6xy^2) + (2x^2y - xy + 2xy^2)$

       $= (4x^2y + 2x^2y) + (-3xy - xy) + (-6xy^2 + 2xy^2) = 6x^2y - 4xy - 4xy^2$

6.6 $(4x^2y - 3xy - 6xy^2) - (2x^2y - xy + 2xy^2)$

       $= (4x^2y - 2x^2y) + (-3xy - (-xy)) + (-6xy^2 - 2xy^2)$

       $= 2x^2y - 2xy - 8xy^2$

6.7 $-3x^2y^3\left(4y^2 - 2y + 6\right) = -12x^2y^5 + 6x^2y^4 - 18x^2y^3$

6.8 $\left(5x + 2\right)\left(2x^2 + 3x - 1\right) = 5x\left(2x^2 + 3x - 1\right) + 2\left(2x^2 + 3x - 1\right)$

       $= 10x^3 + 15x^2 - 5x + 4x^2 + 6x - 2 = 10x^3 + 19x^2 + x - 2$

6.9

$$2y-5\overline{)4y^2-6y-9} \\ \quad 2y+2+\dfrac{1}{2y-5}$$

$$-\left(4y^2-10y\right)$$

$$4y-9$$

$$-\left(4y-10\right)$$

$$1$$

$$\left(4y^2-6y-9\right)\div\left(2y-5\right)=2y+2+\dfrac{1}{2y-5}$$

6.10

$$x+3\overline{)-2x^3+0x^2+13x-15} \\ \quad\ -2x^2+6x-5$$

$$-\left(-2x^3-6x^2\right)$$

$$6x^2+13x$$

$$-\left(6x^2+18x\right)$$

$$-5x-15$$

$$-(-5x-15)$$

$$0$$

$$\left(13x-2x^3-15\right)\div\left(x+3\right)=-2x^2+6x-5$$

6.11

$$\underline{2}|\quad 3 \quad 2 \quad -6 \quad -20$$
$$\qquad\quad 6 \quad\ 16 \quad\ 20$$
$$\overline{\qquad 3 \quad 8 \quad 10 \quad\ \ 0}$$

$$\left(3x^3+2x^2-6x-20\right)\div\left(x-2\right)=3x^2+8x+10$$

6.12

$$\underline{-5}|\quad 1 \quad 6 \quad 3 \quad -8$$
$$\qquad\qquad -5 \quad -5 \quad 10$$
$$\overline{\qquad\ 1 \quad 1 \quad -2 \quad 2}$$

$$\left(x^3+6x^2+3x-8\right)\div\left(x+5\right)=x^2+x-2+\dfrac{2}{x+5}$$

6.13 $\dfrac{\left(6x^4+7x^3+2x-1\right)\div 3}{\left(3x-1\right)\div 3}=\dfrac{2x^4+\dfrac{7}{3}x^3+\dfrac{2}{3}x-\dfrac{1}{3}}{x-\dfrac{1}{3}};$

the divisor is $x-\dfrac{1}{3}$, so $r=\dfrac{1}{3}$.

$$
\begin{array}{r|rrrrr}
\frac{1}{3} & 2 & \frac{7}{3} & 0 & \frac{2}{3} & -\frac{1}{3}\\
 & & \frac{2}{3} & 1 & \frac{1}{3} & \frac{1}{3}\\
\hline
 & 2 & 3 & 1 & 1 & 0
\end{array}
$$

$\left(6x^4+7x^3+2x-1\right)\div\left(3x-1\right)=2x^3+3x^2+x+1$

6.14 $192x^5-81x^2=3x^2(64x^3-27)=3x^2[(4x)^3-3^3]$

$=3x^2(4x-3)(16x^2+12x+9)$

6.15 $x^4-81=\left(x^2\right)^2-81=u^2-81=(u+9)(u-9)=(x^2+9)(x^2-9)$

$=(x^2+9)(x+3)(x-3)$

6.16 $4xy+8xz+6y+12z=4x(y+2z)+6(y+2z)=(4x+6)(y+2z)$

6.17 $x^4+6x^2-16=0;\ \left(x^2\right)^2+6x^2-16=0;\ u^2+6u-16=0;$

$(u+8)(u-2)=0;\ (x^2+8)(x^2-2)=0;\ x^2+8=0$ or $x^2-2=0;$

$x^2=-8$ or $x^2=2;\ x=\pm\sqrt{-8}$ or $x=\pm\sqrt{2};\ x=\pm2i\sqrt{2}$ or $x=\pm\sqrt{2}$

Check $x=2i\sqrt{2}:\ \left(2i\sqrt{2}\right)^4+6\left(2i\sqrt{2}\right)^2-16=16i^4\left(\sqrt{2}\right)^4+$

$6\left(4i^2\left(\sqrt{2}\right)^2\right)-16=64-6\cdot 8-16=0\ \checkmark$ Check $x=-2i\sqrt{2}:$

$\left(-2i\sqrt{2}\right)^4+6\left(-2i\sqrt{2}\right)^2-16=16i^4\left(\sqrt{2}\right)^4+6\left(4i^2\left(\sqrt{2}\right)^2\right)-16=$

$64-6\cdot 8-16=0\ \checkmark$ Check $x=\sqrt{2}:$

$$\left(\sqrt{2}\right)^4 + 6\left(\sqrt{2}\right)^2 - 16 = 4 + 6 \cdot 2 - 16 = 4 + 12 - 16 = 0 \checkmark$$

Check $x = -\sqrt{2}$:

$$\left(-\sqrt{2}\right)^4 + 6\left(-\sqrt{2}\right)^2 - 16 = 4 + 6 \cdot 2 - 16 = 4 + 12 - 16 = 0 \checkmark$$

The solutions are $x = \pm 2i\sqrt{2},\ \pm\sqrt{2}$

6.18 $x^3 - 125 = 0$; $x^3 - 5^3 = 0$; $(x - 5)(x^2 + 5x + 25) = 0$;

$x - 5 = 0$ or $x^2 + 5x + 25 = 0$; $x = 5$ or

$$x = \frac{-5 \pm \sqrt{5^2 - 4(1)(25)}}{2 \cdot 1} = \frac{-5 \pm \sqrt{-75}}{2} = \frac{-5 \pm 5i\sqrt{3}}{2}$$

Check $x = 5$: $5^3 - 125 = 125 - 125 = 0$ $\checkmark$; Check: $x = \dfrac{-5 + 5i\sqrt{3}}{2}$;

$$\left(\frac{-5 + 5i\sqrt{3}}{2}\right)^3 - 125 = 0; \quad \left(\frac{-5 + 5i\sqrt{3}}{2}\right)^2\left(\frac{-5 + 5i\sqrt{3}}{2}\right) - 125 = 0;$$

$$\left(\frac{25 - 50i\sqrt{3} + 25i^2\sqrt{3}^2}{4}\right)\left(\frac{-5 + 5i\sqrt{3}}{2}\right) - 125 = 0;$$

$$\left(\frac{25 - 50i\sqrt{3} - 75}{4}\right)\left(\frac{-5 + 5i\sqrt{3}}{2}\right) - 125 = 0;$$

$$\left(\frac{-50 - 50i\sqrt{3}}{4}\right)\left(\frac{-5 + 5i\sqrt{3}}{2}\right) - 125 = 0;$$

$$\left(\frac{250 - 250i\sqrt{3} + 250i\sqrt{3} - 250i^2\sqrt{3}^2}{8}\right) - 125 = 0;$$

$$\left(\frac{250 + 750}{8}\right) - 125 = 0; \quad 125 - 125 = 0 \checkmark \quad \text{The check for}$$

$x = \dfrac{-5 - 5i\sqrt{3}}{2}$ is left to you. The solutions are $x = 5, \dfrac{-5 \pm 5i\sqrt{3}}{2}$.

6.19 $P(4) = 4^2 - 3(4) + 4 = 16 - 12 + 4 = 8$

6.20 $P(x - h) = (x - h)^2 - 3(x - h) + 4$

$P(x - h) = x^2 - 2xh + h^2 - 3x + 3h + 4$

6.21   The function can have at most four zeros. $x^4 + 2x^2 - 3 = 0$;

$\left(x^2\right)^2 + 2\left(x^2\right) - 3 = 0$; $u^2 + 2u - 3 = 0$; $(u + 3)(u - 1) = 0$;

$(x^2 + 3)(x^2 - 1) = 0$; $x^2 + 3 = 0$ or $x^2 - 1 = 0$; $x^2 = -3$ or $x^2 = 1$;

$x = \pm\sqrt{-3}$ or $x = \pm 1$. The zeros of the

function are $\pm\sqrt{-3}$ and ± 1.

The only real zeros of the function are
± 1. The points $(-1, 0)$ and $(1, 0)$ are on
the graph.

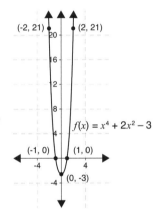

x	$x^4 + 2x^2 - 3$	$f(x)$
–2	$(-2)^4 + 2(-2)^2 - 3$	21
0	$(0)^4 + 2(0)^2 - 3$	–3
2	$(2)^4 + 2(2)^2 - 3$	21

6.22   The degree of the function is 3, and the leading coefficient
is positive, so the left end of the graph will go down and the
right end of the graph will go up. The polynomial is not readily
factorable, so make a table of values. The maximum number of
real zeros is 3 since the degree of the function is 3.

x	$4x^3 + 4x^2 - 11x - 6$	$f(x)$	
–3	$4(-3)^3 + 4(-3)^2 - 11(-3) - 6$	–45	
–2	$4(-2)^3 + 4(-2)^2 - 11(-2) - 6$	0	–2 is a zero of $f(x)$
–1	$4(-1)^3 + 4(-1)^2 - 11(-1) - 6$	5	
0	$4(0)^3 + 4(0)^2 - 11(0) - 6$	–6	$f(x)$ changes sign between $x = -1, 0$
1	$4(1)^3 + 4(1)^2 - 11(1) - 6$	–9	
2	$4(2)^3 + 4(2)^2 - 11(2) - 6$	20	$f(x)$ changes sign between $x = 1, 2$

There is a zero at –2, between –1 and 0, and between 1 and 2. The values of $f(x)$ are increasing to negative infinity on the left and positive infinity on the right. Using the values from the table and the end behavior of the graph, make a sketch of the graph.

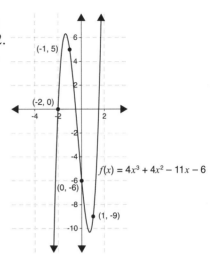

$f(x) = 4x^3 + 4x^2 - 11x - 6$

(-1, 5)
(-2, 0)
(0, -6)
(1, -9)

6.23 $P(-2) = 3(-2)^4 - 2(-2)^3 + 4(-2) - 2 = 3(16) - 2(-8) + 4(-2) - 2 = 54$

6.24 $\underline{-2|}\ 3\ -2\quad 0\quad 4\quad -2$
       $\qquad\ \ \underline{-6\quad 16\ -32\quad 56}$
       $\qquad\ \ 3\ -8\quad 16\ -28\quad 54$ The remainder is 54.

6.25 $\underline{1|}\ 1\ -1\ -4\quad 4$
       $\qquad\ \underline{\ \ 1\quad 0\ -4}$
       $\qquad\ 1\quad 0\ -4\quad 0$ Since the remainder is 0, $x - 1$ is a factor

and 1 is a zero of $P(x)$. The synthetic division shows that

$(x^3 - x^2 - 4x + 4) \div (x - 1) = x^2 - 4$. $x^2 - 4 = (x + 2)(x - 2)$, so

$x^3 - x^2 - 4x + 4 = (x - 1)(x + 2)(x - 2)$. The factors of $P(x)$ are

$(x - 1)$, $(x + 2)$, and $(x - 2)$. The zeros of $P(x)$ are 1, –2, and 2.

6.26 $\underline{1|}\ 1\ -5\ -8\quad 12$
       $\qquad\ \underline{\ \ 1\ -4\ -12}$
       $\qquad\ 1\ -4\ -12\quad 0$ Since the remainder is 0, $x - 1$ is a factor

and 1 is a zero of $P(x)$. The synthetic division shows that

$(x^3 - 5x^2 - 8x + 12) \div (x - 1) = x^2 - 4x - 12$.

$x^2 - 4x - 12 = (x + 2)(x - 6)$ so

$x^3 - 5x^2 - 8x + 12 = (x - 1)(x + 2)(x - 6)$. The factors of $P(x)$ are

$(x - 1)$, $(x + 2)$, and $(x - 6)$. The zeros of $P(x)$ are 1, –2, and 6.

6.27 p: factors of -6: ± 1, ± 2, ± 3, ± 6; q: factors of 3: ± 1, ± 3

$\dfrac{p}{q}$: ± 1, ± 2, ± 3, ± 6, $\pm\dfrac{1}{3}$, $\pm\dfrac{2}{3}$, $\pm\dfrac{3}{2}$; The possible rational

zeros of $P(x)$ are ± 1, ± 2, ± 3, ± 6, $\pm\dfrac{1}{3}$, $\pm\dfrac{1}{2}$, $\pm\dfrac{2}{3}$, $\pm\dfrac{3}{2}$.

6.28

$\dfrac{p}{q}$	3	4	−17	−6	
−1	3	1	−18	12	
1	3	7	−10	−16	
−2	3	−2	−13	20	
2	3	10	3	0	2 is a zero.
−3	3	−5	−2	0	−3 is a zero.
3	3	13	9	21	
−6	3	−14	67	−408	
6	3	22	115	684	
$-\dfrac{1}{3}$	3	3	−18	0	$-\dfrac{1}{3}$ is a zero.

The rational zeros of $P(x)$ are -3, $-\dfrac{1}{3}$, and 2.

6.29   Since the leading coefficient is 1, the possible rational zeros are the factors of p. Factors of p: ± 1, ± 2, ± 3, ± 4, ± 6, ± 8, ± 12, ± 16, ± 24, ± 48.

$\dfrac{p}{q}$	1	0	−28	48	
−1	1	−1	−27	75	
1	1	1	−27	21	
−2	1	−2	−24	96	
2	1	2	−24	0	2 is a zero.

The synthetic division shows that one rational zero of $P(x)$ is 2 and the remaining factor is $x^2 + 2x - 24$. $x^2 + 2x - 24 = 0$; $(x+6)(x-4) = 0$; $x + 6 = 0$ or $x - 4 = 0$; $x = -6$ or $x = 4$; The zeros of the function are 2, 4, and -6.

6.30 Since the leading coefficient is 1, the possible rational solutions are the factors of p.

Factors of p: $\pm1, \pm2, \pm4$

$\dfrac{p}{q}$	1	-3	-5	4
-1	1	-4	-1	5
1	1	-2	-7	-3
-2	1	-5	5	-6
2	1	-1	-7	-10
-4	1	-7	23	-88
4	1	1	-1	0

The synthetic division shows that 4 is a solution and $x^3 - 3x^2 - 5x + 4 = (x-4)(x^2 + x - 1)$. Solve the depressed equation

$x^2 + x - 1 = 0$ to find the remaining solutions. $x^2 + x - 1 = 0$; $a = 1$,

$b = 1, c = -1$; $x = \dfrac{-1 \pm \sqrt{1^2 - 4(1)(-1)}}{2(1)} = \dfrac{-1 \pm \sqrt{5}}{2}$;

The solutions are 4, $\dfrac{-1 \pm \sqrt{5}}{2}$.

Check: $x = 4$: $(4)^3 - 3(4)^2 - 5(4) + 4 = 64 - 48 - 20 + 4 = 0$ ✓

Check $x = \dfrac{-1 + \sqrt{5}}{2} \approx 0.61803$:

$(0.61803)^3 - 3(0.61803)^2 - 5(0.61803) + 4 \approx 0$ ✓

Check $x = \dfrac{-1 - \sqrt{5}}{2} \approx -1.61803$:

$(-1.61803)^3 - 3(-1.61803)^2 - 5(-1.61803) + 4 \approx 0$ ✓

6.31 $P(x) = x^3 - 3x^2 - 5x + 4$

$$\underbrace{+ \quad -}_{1} \quad \underbrace{- \quad +}_{2} \qquad \text{2 sign changes}$$

By Descartes' Rule, $P(x)$ can have two or no positive real zeros.

$P(-x) = (-x)^3 - 3(-x)^2 - 5(-x) + 4 \qquad P(-x) = -x^3 - 3x^2 + 5x + 4$

$$- \quad - \quad \underbrace{+ \quad +}_{1}$$

1 sign change

By Descartes' Rule, $P(x)$ has one negative real zero.

$P(x)$ has one negative real zero and two or no positive real zeros.

6.32

x	1	−3	−5	4	
1	1	−2	−7	−3	
2	1	−1	−7	−10	
3	1	0	−5	−11	
4	1	1	−1	0	4 is a zero of the function.
5	1	2	5	29	Remainder and coefficients all positive.

$P(-x) = (-x)^3 - 3(-x)^2 - 5(-x) + 4 \,; \; P(-x) = -x^3 - 3x^2 + 5x + 4$

x	1	−3	−5	4	
1	−1	−4	1	5	
2	−1	−5	−5	−6	Remainder and coefficients all negative.

so −2 is a lower bound of $P(x)$. A lower bound of $P(x)$ is −2 and an upper bound is 5. Note also that 4 is a zero of the function.

6.33   Exercise 6.32 showed that the zeros of the function will occur between −2 and 5 and that 4 is a zero of the function. Exercise 6.31 showed that $P(x)$ has one negative real zero and two or no positive real zeros. Since we have found a positive zero, there must be two positive zeros rather than no positive zeros. Use synthetic division to test values between −2 and 0 to locate one negative real zero and between 0 and 5 to locate one positive real zero. A change of sign in the remainder between two successive integers indicates that at least one zero lies between the two integers.

x	1	−3	−5	4
−2	1	−5	5	−6
−1	1	−4	−1	5

−1 ... Sign change: zero between $x = -2$ and −1.

There is only one negative real zero, and it lies between −2 and −1.

x	1	−3	−5	4
0	1	−3	−5	4
1	1	−2	−7	−3

1 ... Sign change: zero between $x = 0$ and 1.

There is a zero between 0 and 1, and a zero between −2 and −1.

6.34 Step 1: Determine the number of zeros. The degree of $P(x)$ is 4, so there are at most four zeros. Step 2: Use Descartes' Rule to determine the possible number of positive and negative real zeros.

$P(x) = x^4 + 3x^3 + 4x^2 - 8$ There is one positive real zero.
$\quad\quad\quad + \quad + \quad + \quad -$

1   1 sign change

$P(-x) = (-x)^4 + 3(-x)^3 + 4(-x)^2 - 8$

$P(-x) = x^4 - 3x^3 + 4x^2 - 8$ There are three or one negative
$\quad\quad\quad + \quad - \quad + \quad -$ real zeros.

1 2 3   3 sign changes

Step 3: Evaluate the upper and lower bounds.

$P(x) = x^4 + 3x^3 + 4x^2 - 8$

x	1	3	4	0	−8
1	1	4	8	8	0

1 is a zero of $P(x)$.

There is only one positive real zero, and it is 1. An upper bound is 1.

$P(-x) = x^4 - 3x^3 + 4x^2 - 8$

x	1	−3	4	0	−8
1	1	−2	2	2	−6
2	1	−1	2	4	0
3	1	0	4	12	28

Remainder and coefficients all positive.

A lower bound is −3.

Step 4: Locate a zero. When testing for the upper bound, 1 was identified as a zero.

Step 5: Write the depressed equation and locate a zero.

From the synthetic division table, when $x = 1$, the depressed equation is $x^3 + 4x^2 + 8x + 8 = 0$. $p = \pm 1, \pm 2, \pm 4, \pm 8$

The only positive zero has already been found. A lower bound is -3.

Remaining values of $p = -1, -2$

$x^3 + 4x^2 + 8x + 8 = 0$

p	1	4	8	8	
-1	1	3	5	3	
-2	1	2	4	0	-2 is a zero of $P(x)$.

Step 6: Continue until the depressed equation is a quadratic equation. Solve using quadratic equation techniques. From the last line of the synthetic division, the depressed equation is

$$x^2 + 2x + 4 = 0; \ a = 1, \ b = 2, \ c = 4; \ x = \frac{-2 \pm \sqrt{(2)^2 - 4(1)(4)}}{2(1)}$$

$$= \frac{-2 \pm \sqrt{-12}}{2} = \frac{-2 \pm 2i\sqrt{3}}{2} = -1 \pm i\sqrt{3}; \ -1 \pm i\sqrt{3} \text{ are zeros of } P(x).$$

The zeros of $P(x)$ are $1, -2, -1 + i\sqrt{3}$; and $-1 - i\sqrt{3}$.

Chapter 7

7.1   Domain: $x + 1 \geq 0$; $x \geq -1$; Range: $f(-1) = 2\sqrt{-1+1} = 2\sqrt{0} = 0$; $f(x) \geq 0$

7.2   Domain: $x - 2 \geq 0$; $x \geq 2$; Range: $f(2) = \sqrt{2-2} - 3 = -3$; $f(x) \geq -3$

7.3 $h = -1$; the function is translated 1 unit to the left; $a = 2$; the function is stretched vertically; $k = 0$; Domain: $x \geq -1$; Range: $f(x) \geq 0$

x	$2\sqrt{x+1}$	$f(x)$
-1	$2\sqrt{-1+1}$	0
0	$2\sqrt{0+1}$	2
3	$2\sqrt{3+1}$	4

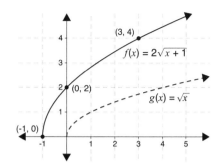

7.4 $h = 2$; the function is translated 2 units to the right; $a = 1$; the function is not reflected or dilated; $k = -3$; the function is translated 3 units down; Domain: $x \geq 2$; Range: $f(x) \geq -3$

x	$\sqrt{x-2}-3$	$f(x)$
2	$\sqrt{2-2}-3$	-3
3	$\sqrt{3-2}-3$	-2
6	$\sqrt{6-2}-3$	-1

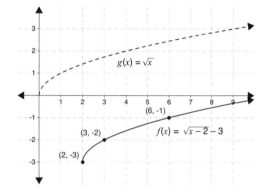

7.5 $\sqrt[3]{64} = \sqrt[3]{4^3} = 4$

7.6   Not real; complex: $2i$

7.7 $\sqrt[3]{-64} = \sqrt[3]{(-4)^3} = -4$

7.8 $\sqrt[4]{256} = \sqrt[4]{4^4} = 4$

7.9 $\sqrt{16x^2y^4z^2} = \sqrt{4^2x^2\left(y^2\right)^2z^2} = 4|x|y^2|z|$

7.10 $\sqrt[3]{27x^3y^6} = \sqrt[3]{3^3x^3\left(y^2\right)^3} = 3xy^2$

7.11 $\sqrt[4]{48x^5y^2z^4} = \sqrt[4]{2^4\cdot3x^4xy^2z^4} = \sqrt[4]{2^4x^4z^4}\sqrt[4]{3xy^2} = 2x|z|\sqrt[4]{3xy^2}$

7.12 $\sqrt[3]{-54x^4y^3z} = \sqrt[3]{(-3)^3\cdot2x^3xy^3z} = \sqrt[3]{(-3)^3x^3y^3}\sqrt[3]{2xz} = -3xy\sqrt[3]{2xz}$

7.13 $6\sqrt{2} - \sqrt{8} + \sqrt{50} = 6\sqrt{2} - \sqrt{2^2\cdot2} + \sqrt{5^2\cdot2} = 6\sqrt{2} - 2\sqrt{2} + 5\sqrt{2} = 9\sqrt{2}$

7.14 $\sqrt{20}+\sqrt{45}-3\sqrt{5}=\sqrt{2^2\cdot5}+\sqrt{3^2\cdot5}-3\sqrt{5}=2\sqrt{5}+3\sqrt{5}-3\sqrt{5}=2\sqrt{5}$

7.15 $\left(\sqrt{3}+1\right)\left(\sqrt{6}-2\right)=\sqrt{3}\cdot\sqrt{6}-2\sqrt{3}+\sqrt{6}-2=\sqrt{18}-2\sqrt{3}+\sqrt{6}-2$

$=\sqrt{3^2\cdot2}-2\sqrt{3}+\sqrt{6}-2=3\sqrt{2}-2\sqrt{3}+\sqrt{6}-2$

7.16 $\left(\sqrt{2}+\sqrt{10}\right)^2=\left(\sqrt{2}\right)^2+2\sqrt{2}\cdot\sqrt{10}+\left(\sqrt{10}\right)^2=2+2\sqrt{20}+10$

$=12+2\sqrt{2^2\cdot5}=12+4\sqrt{5}$

7.17 $\sqrt[4]{\dfrac{x}{81}}=\dfrac{\sqrt[4]{x}}{\sqrt[4]{81}}=\dfrac{\sqrt[4]{x}}{\sqrt[4]{3^4}}=\dfrac{\sqrt[4]{x}}{3}$

7.18 $\dfrac{\sqrt[3]{40}}{\sqrt[3]{5}}=\sqrt[3]{\dfrac{40}{5}}=\sqrt[3]{8}=\sqrt[3]{2^3}=2$

7.19 $\sqrt[3]{\dfrac{8}{x}}=\dfrac{\sqrt[3]{8}}{\sqrt[3]{x}}=\dfrac{\sqrt[3]{2^3}}{\sqrt[3]{x}}=\dfrac{2}{\sqrt[3]{x}}\cdot\dfrac{\sqrt[3]{x^2}}{\sqrt[3]{x^2}}=\dfrac{2\sqrt[3]{x^2}}{\sqrt[3]{x^3}}=\dfrac{2\sqrt[3]{x^2}}{x}$

7.20 $\sqrt{\dfrac{16x^4}{7y}}=\dfrac{\sqrt{4^2\left(x^2\right)^2}}{\sqrt{7y}}=\dfrac{4x^2}{\sqrt{7y}}\cdot\dfrac{\sqrt{7y}}{\sqrt{7y}}=\dfrac{4x^2\sqrt{7y}}{\left(\sqrt{7y}\right)^2}=\dfrac{4x^2\sqrt{7y}}{7y}$

7.21 $4\sqrt{2}-2\sqrt{5}$

7.22 $\dfrac{\sqrt{2}+3}{\sqrt{3}-\sqrt{2}}\cdot\dfrac{\sqrt{3}+\sqrt{2}}{\sqrt{3}+\sqrt{2}}=\dfrac{\sqrt{6}+\sqrt{2}\cdot\sqrt{2}+3\sqrt{3}+3\sqrt{2}}{\left(\sqrt{3}\right)^2-\left(\sqrt{2}\right)^2}$

$=\dfrac{\sqrt{6}+\sqrt{4}+3\sqrt{3}+3\sqrt{2}}{3-2}=\sqrt{6}+2+3\sqrt{3}+3\sqrt{2}$

7.23 $\left(16\right)^{-\frac{3}{4}}=\dfrac{1}{16^{\frac{3}{4}}}=\dfrac{1}{\left(\sqrt[4]{16}\right)^3}=\dfrac{1}{\left(\sqrt[4]{2^4}\right)^3}=\dfrac{1}{2^3}=\dfrac{1}{8}$

7.24 $\left(-27x^4y^5\right)^{\frac{1}{3}}=\sqrt[3]{-27x^4y^5}=\sqrt[3]{\left(-3\right)^3x^3xy^3y^2}=-3xy\sqrt[3]{xy^2}$

7.25 $x^{\frac{2}{3}}\cdot x^{\frac{1}{4}}=x^{\frac{2}{3}+\frac{1}{4}}=x^{\frac{8}{12}+\frac{3}{12}}=x^{\frac{11}{12}}$

7.26 $\quad x^{-\frac{3}{5}} = \dfrac{1}{x^{\frac{3}{5}}} \cdot \dfrac{x^{\frac{2}{5}}}{x^{\frac{2}{5}}} = \dfrac{x^{\frac{2}{5}}}{x^{\frac{3}{5}} \cdot x^{\frac{2}{5}}} = \dfrac{x^{\frac{2}{5}}}{x^{\frac{3}{5}+\frac{2}{5}}} = \dfrac{x^{\frac{2}{5}}}{x}$

7.27 $\quad \dfrac{\sqrt{3}}{\sqrt[5]{9}} = \dfrac{3^{\frac{1}{2}}}{9^{\frac{1}{5}}} = \dfrac{3^{\frac{1}{2}}}{\left(3^2\right)^{\frac{1}{5}}} = \dfrac{3^{\frac{1}{2}}}{3^{\frac{2}{5}}} = 3^{\frac{1}{2}-\frac{2}{5}} = 3^{\frac{5}{10}-\frac{4}{10}} = 3^{\frac{1}{10}} = \sqrt[10]{3}$

7.28 $\quad \sqrt{(xy)^3} \cdot \sqrt[4]{xy} = (xy)^{\frac{3}{2}}(xy)^{\frac{1}{4}} = (xy)^{\frac{3}{2}+\frac{1}{4}} = (xy)^{\frac{6}{4}+\frac{1}{4}} = (xy)^{\frac{7}{4}}$

$\qquad = \sqrt[4]{(xy)^7} = \sqrt[4]{(xy)^4(xy)^3} = xy\sqrt[4]{(xy)^3}$

7.29 $\quad -2\sqrt[3]{2x+10} = 4;\ \sqrt[3]{2x+10} = -2;\ \left(\sqrt[3]{2x+10}\right)^3 = (-2)^3;\ 2x+10 = -8;$

$\qquad 2x = -18;\ x = -9;$ Check $x = -9$: $-2\sqrt[3]{2(-9)+10} = -2(-2) = 4$ ✓
The solution is -9.

7.30 $\quad \sqrt{x+3} = x+3;\ \left(\sqrt{x+3}\right)^2 = (x+3)^2;\ x+3 = x^2+6x+9;$

$\qquad 0 = x^2+5x+6;\ 0 = (x+3)(x+2);\ x = -3, -2;$ Check $x = -3$:

$\qquad \sqrt{-3+3} = -3+3;\ 0 = 0$ ✓ Check $x = -2$. $\sqrt{-2+3} = -2+3;$

$\qquad 1 = 1$ ✓ The solutions are -2 and -3.

7.31 $\quad \sqrt{2x+5} = \sqrt{x+2}+1;\ \left(\sqrt{2x+5}\right)^2 = \left(\sqrt{x+2}+1\right)^2;$

$\qquad 2x+5 = \left(\sqrt{x+2}\right)^2 + 2\sqrt{x+2} + 1;\ 2x+5 = x+2+2\sqrt{x+2}+1;$

$\qquad 2x+5 = x+3+2\sqrt{x+2};\ x+2 = 2\sqrt{x+2};\ (x+2)^2 = \left(2\sqrt{x+2}\right)^2;$

$\qquad x^2+4x+4 = 4(x+2);\ x^2+4x+4 = 4x+8;\ x^2-4 = 0;$

$\qquad (x+2)(x-2) = 0;\ x = -2, 2;$ Check $x = -2$: $\sqrt{2(-2)+5} = \sqrt{-2+2}+1;$

$\qquad \sqrt{1} = \sqrt{0}+1;\ 1 = 1$ ✓ Check $x = 2$: $\sqrt{2(2)+5} = \sqrt{2+2}+1;$

$\qquad \sqrt{9} = \sqrt{4}+1;\ 3 = 2+1$ ✓ The solutions are -2 and 2.

7.32 $(3x+2)^{\frac{2}{5}} = 4$; $\left[(3x+2)^{\frac{2}{5}}\right]^{\frac{5}{2}} = 4^{\frac{5}{2}}$; $3x+2 = \left(\sqrt{4}\right)^5$; $3x+2 = 2^5$;

3x + 2 = 32; 3x = 30; x = 10. Check: x = 10; $(3(10)+2)^{\frac{2}{5}} = 4$;

$(32)^{\frac{2}{5}} = 4$; $\left(\sqrt[5]{32}\right)^2 = 4$; 4 = 4 ✓ The solution is 10.

Chapter 8

8.1

x	3^x	f(x)
−2	$3^{-2} = \dfrac{1}{3^2}$	$\dfrac{1}{9}$
−1	$3^{-1} = \dfrac{1}{3^1}$	$\dfrac{1}{3}$
0	3^0	1
1	3^1	3
2	3^2	9

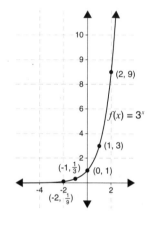

8.2

x	$\left(\dfrac{2}{3}\right)^x$	g(x)
−2	$\left(\dfrac{2}{3}\right)^{-2} = \left(\dfrac{3}{2}\right)^2$	$\dfrac{9}{4}$
−1	$\left(\dfrac{2}{3}\right)^{-1} = \left(\dfrac{3}{2}\right)^1$	$\dfrac{3}{2}$
0	$\left(\dfrac{2}{3}\right)^0$	1
1	$\left(\dfrac{2}{3}\right)^1$	$\dfrac{2}{3}$
2	$\left(\dfrac{2}{3}\right)^2$	$\dfrac{4}{9}$

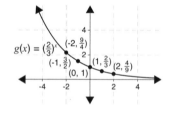

8.3 $h = 1$, $a = 1$, and $k = 0$. The graph of the parent function is
       translated 1 unit to the right. The asymptote is the line $y = 0$.

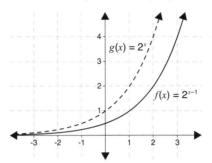

8.4 $h = -4$, $a = 3$, and $k = -1$. The graph of the parent function is
       translated 4 units to the left, stretched vertically, and translated
       one unit down. The asymptote is the line $y = -1$.

x	$3(2)^{x+4} - 1$	$f(x)$
-4	$3(2)^{-4+4} - 1$	2
-3	$3(2)^{-3+4} - 1$	5
-2	$3(2)^{-2+4} - 1$	11

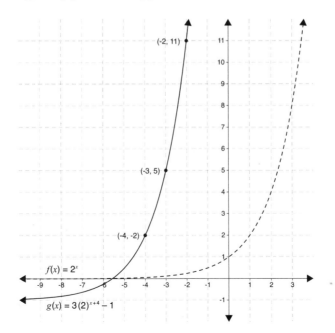

8.5 $a = \$5,000;\ r = 6\% = 0.06;\ A(t) = \$5,000(1+0.06)^t$

      $A(t) = \$5,000(1.06)^t;\ \ A(1) = \$5,000(1.06)^1 = \$5,300$

      $A(2) = \$5,000(1.06)^2 = \$5,618;\ \ A(3) = \$5,000(1.06)^3 = \$5,955.08$

      $A(4) = \$5,000(1.06)^4 = \$6,312.38$

      $A(5) = \$5,000(1.06)^5 = \$6,691.13$

8.6 $a = 105,000;\ r = -2.5\% = -0.025;\ A(t) = 105,000(1-0.025)^t;$

      $A(t) = 105,000(0.975)^t;\ \ A(10) = 105,000(0.975)^{10} = 81,515$

      If the population continues to decline by 2.5% per year, the population in 10 years will be 81,515.

8.7 $1,024 = 4^x;\ 4^5 = 4^x;\ x = 5;$ Check $x = 5$: $1,024 = 4^5;\ 1,024 = 1,024$ ✓

8.8 $27 = 3^{4x+1};\ 3^3 = 3^{4x+1};\ 3 = 4x + 1;\ 2 = 4x;\ \dfrac{1}{2} = x;$ Check $x = \dfrac{1}{2}$:

      $27 = 3^{4\left(\frac{1}{2}\right)+1};\ 27 = 3^{2+1};\ 27 = 3^3;\ 27 = 27$ ✓

8.9 $64^{2x-3} = 16^{x-2};\ 4^{3(2x-3)} = 4^{2(x-2)};\ 3(2x - 3) = 2(x - 2);$

      $6x - 9 = 2x - 4;\ 4x = 5;\ x = \dfrac{5}{4};$ Check $x = \dfrac{5}{4}$: $64^{2\left(\frac{5}{4}\right)-3} = 16^{\frac{5}{4}-2};$

      $64^{\frac{5}{2}-3} = 16^{-\frac{3}{4}};\ 64^{-\frac{1}{2}} = 16^{-\frac{3}{4}};\ \dfrac{1}{64^{\frac{1}{2}}} = \dfrac{1}{16^{\frac{3}{4}}}\ ;\ \dfrac{1}{\sqrt{64}} = \dfrac{1}{\sqrt[4]{16}^3};\ \dfrac{1}{8} = \dfrac{1}{8}$ ✓

8.10 $\log_4 64 = 3$

8.11 $2^5 = 32$

8.12 $x = \log_6 36;\ 6^x = 36;\ 6^x = 6^2;\ x = 2;\ \log_6 36 = 2$

8.13 $\log_x 125 = 3;\ x^3 = 125;\ x^3 = 5^3;\ \left(x^3\right)^{\frac{1}{3}} = \left(5^3\right)^{\frac{1}{3}};\ x = 5$

      Check $x = 5$: $\log_5 125 = 3;\ 5^3 = 125;\ 125 = 125$ ✓

8.14 $\log_6 x = 3$; $6^3 = x$; $x = 216$; Check $x = 216$: $\log_6 216 = 3$;

$6^3 = 216$; $216 = 216$ ✓

8.15 $\log_5 625 = 2x + 1$; $5^{2x+1} = 625$; $5^{2x+1} = 5^4$; $2x + 1 = 4$; $2x = 3$; $x = \dfrac{3}{2}$

Check $x = \dfrac{3}{2}$: $\log_5 625 = 2\left(\dfrac{3}{2}\right) + 1$; $\log_5 625 = 4$; $5^4 = 625$;

$625 = 625$ ✓

8.16 $\log_2 \dfrac{x^2}{4} = \log_2 x^2 - \log_2 4 = 2\log_2 x - \log_2 4 = 2\log_2 x - 2$

8.17 $\log_2 72 = \log_2 6^2 \cdot 2 = \log_2 6^2 + \log_2 2 = \log_2 6^2 + 1 = 2\log_2 6 + 1$

$\approx 2(2.585) + 1$; ≈ 6.17

8.18 $\log_4 x = 5\log_4 2 - \log_4 8$; $\log_4 x = \log_4 2^5 - \log_4 8$; $\log_4 x = \log_4 \dfrac{32}{8}$;

$\log_4 x = \log_4 4$; $x = 4$; Check $x = 4$: $\log_4 4 = 5\log_4 2 - \log_4 8$;

$\log_4 4 = \log_4 \dfrac{2^5}{8}$; $\log_4 4 = \log_4 4$ ✓

8.19 $\log_6 x + \log_6 18 = 3$; $\log_6 18x = 3$; $6^3 = 18x$; $216 = 18x$; $12 = x$;

Check $x = 12$: $\log_6 12 + \log_6 18 = 3$; $\log_6 12 \cdot 18 = 3$; $\log_6 216 = 3$;

$6^3 = 216$; $216 = 216$ ✓

8.20 $\log_3 x + \log_3(x + 4) = 2$; $\log_3 x(x + 4) = 2$; $3^2 = x(x + 4)$;

$9 = x^2 + 4x$; $0 = x^2 + 4x - 9$; $a = 1$, $b = 4$, $c = -9$;

$x = \dfrac{-4 \pm \sqrt{4^2 - 4(1)(-9)}}{2 \cdot 1}$; $x = \dfrac{-4 \pm \sqrt{52}}{2}$; $x = \dfrac{-4 \pm \sqrt{13 \cdot 4}}{2}$;

$x = \dfrac{-4 \pm 2\sqrt{13}}{2} = -2 \pm \sqrt{13}$

Check $x = -2 + \sqrt{13}$: $\log_3\left(-2 + \sqrt{13}\right) + \log_3\left(-2 + \sqrt{13} + 4\right) = 2$;

$\log_3\left(-2 + \sqrt{13}\right) + \log_3\left(2 + \sqrt{13}\right) = 2$;

$\log_3\left[\left(-2 + \sqrt{13}\right)\left(2 + \sqrt{13}\right)\right] = 2$; $\log_3\left[\left(\sqrt{13} - 2\right)\left(\sqrt{13} + 2\right)\right] = 2$;

$\log_3\left(13 - 4\right) = 2$; $\log_3\left(9\right) = 2$; $3^2 = 9$ ✓ Check $x = -2 - \sqrt{13}$:

$\log_3\left(-2 - \sqrt{13}\right) + \log_3\left(-2 - \sqrt{13} + 4\right) = 2$ Undefined. $x = -2 - \sqrt{13}$

is not a solution; $x = -2 + \sqrt{13} \approx 1.6056$

8.21 $\log_4 7 = \dfrac{\log 7}{\log 4} \approx 1.4037$

8.22 $82 = 4^x$; $\log 82 = \log 4^x$; $\log 82 = x \log 4$; $\dfrac{\log 82}{\log 4} = x$; $x \approx 3.1788$

        Check $x = \dfrac{\log 82}{\log 4} \approx 3.1788$: $82 = 4^{3.1788} \approx 82.0027$ ✓

8.23 $\left(\dfrac{12e^4}{6e^2}\right)^2 = \left(2e^2\right)^2 = 2^2\left(e^2\right)^2 = 4e^4$

8.24 $\ln\sqrt{xy} = \ln\left(xy\right)^{\frac{1}{2}} = \dfrac{1}{2}\ln\left(xy\right) = \dfrac{1}{2}\left(\ln x + \ln y\right)$

8.25 $2\ln 3x - 4\ln y = \ln\left(3x\right)^2 - \ln y^4 = \ln\dfrac{\left(3x\right)^2}{y^4} = \ln\dfrac{9x^2}{y^4}$

8.26 $10^{2x} - 3 = 5$; $10^{2x} = 8$; $\log 10^{2x} = \log 8$; $2x = \log 8$;

        $x = \dfrac{\log 8}{2} \approx 0.4515$; Check $x = \dfrac{\log 8}{2}$: $10^{2\left(\frac{\log 8}{2}\right)} - 3 = 5$;

        $10^{\log 8} - 3 = 5$; $8 - 3 = 5$ ✓

8.27 $3e^{4x-2} + 1 = 7$; $3e^{4x-2} = 6$; $e^{4x-2} = 2$; $\ln e^{4x-2} = \ln 2$; $4x - 2 = \ln 2$;

$4x = 2 + \ln 2$; $x = \dfrac{2 + \ln 2}{4} \approx 0.6733$. Check $x = \dfrac{2 + \ln 2}{4} \approx 0.6733$:

$3e^{4\left(\frac{2+\ln 2}{4}\right)-2} + 1 = 7$; $3e^{2+\ln 2 - 2} + 1 = 7$; $3e^{\ln 2} + 1 = 7$; $3(2) + 1 = 7$;

$6 + 1 = 7$ ✓

8.28 $6\ln 5x^2 - 2 = 4$; $6\ln 5x^2 = 6$; $\ln 5x^2 = 1$; $e^{\ln 5x^2} = e^1$; $5x^2 = e^1$;

$x^2 = \dfrac{e^1}{5}$; $x = \pm\sqrt{\dfrac{e^1}{5}} \approx \pm 0.7373$. Check $x = \sqrt{\dfrac{e^1}{5}}$:

$6\ln 5\left(\sqrt{\dfrac{e}{5}}\right)^2 - 2 = 4$; $6\ln 5\left(\dfrac{e}{5}\right) - 2 = 4$; $6\ln e - 2 = 4$; $6 - 2 = 4$ ✓

Check $x = -\sqrt{\dfrac{e^1}{5}}$: $6\ln 5\left(-\sqrt{\dfrac{e}{5}}\right)^2 - 2 = 4$; $6\ln 5\left(\dfrac{e}{5}\right) - 2 = 4$;

$6\ln e - 2 = 4$; $6 - 2 = 4$ ✓

8.29 $4\log 2x - 3 = 1$; $4\log 2x = 4$; $\log 2x = 1$; $10^{\log 2x} = 10^1$; $2x = 10$;

$x = 5$. Check $x = 5$: $4\log 2(5) - 3 = 1$; $4\log 10 - 3 = 1$

$4(1) - 3 = 1$ ✓

Chapter 9

9.1 $\dfrac{(x-1)^3}{x^2 + x - 2} = \dfrac{(x-1)(x-1)\,\cancel{(x-1)}^{\,1}}{(x+2)\,\cancel{(x-1)}_{\,1}} = \dfrac{(x-1)(x-1)}{x+2}$ or $\dfrac{x^2 - 2x + 1}{x+2}$;

The rational expression would be undefined if $x^2 + x - 2 = 0$.

$x^2 + x - 2 = 0$; $(x + 2)(x - 1) = 0$; $x = -2$ or $x = 1$; So, the rational expression would be undefined for $x = -2$ or 1.

$\dfrac{(x-1)^3}{x^2 + x - 2} = \dfrac{(x-1)(x-1)}{x+2}$ or $\dfrac{x^2 - 2x + 1}{x+2}$ when $x \neq -2$ or 1.

9.2 $\dfrac{(x-3)(x^2+2x-8)}{(4-x^2)(4+x)} = \dfrac{(x-3)(x+4)(x-2)}{(2+x)(2-x)(4+x)} = \dfrac{(x-3)\,\cancel{(x+4)}\,(-1)\,\cancel{(2-x)}}{(2+x)\,\cancel{(2-x)}\,\cancel{(4+x)}} = -\dfrac{x-3}{x+2}$

The rational expression would be undefined if $(4-x^2)(4+x) = 0$.

$(4-x^2)(4+x) = 0$; $(2+x)(2-x)(4+x) = 0$; $x = -2, 2, -4$; So the rational expression would be undefined for $x = -4, -2,$ and 2.

$\dfrac{(x-3)(x^2+2x-8)}{(4-x^2)(4+x)} = -\dfrac{x-3}{x+2}$ when $x \neq -4, -2$ or 2.

9.3 $\dfrac{x^2-3x-28}{x^2-5x-14} \cdot \dfrac{6x^3+12x^2}{x^2+8x+16} = \dfrac{\cancel{(x-7)}\,\cancel{(x+4)}}{\cancel{(x-7)}\,\cancel{(x+2)}} \cdot \dfrac{6x^2\,\cancel{(x+2)}}{\cancel{(x+4)}(x+4)} = \dfrac{6x^2}{x+4}$

9.4 $\dfrac{x+4}{x^2+7x+12} \cdot \dfrac{x+3}{x} = \dfrac{\cancel{x+4}}{\cancel{(x+3)}\,\cancel{(x+4)}} \cdot \dfrac{\cancel{x+3}}{x} = \dfrac{1}{x}$

9.5 $\dfrac{3x^2-12}{4-x} \div \dfrac{x-2}{8x-32} = \dfrac{3x^2-12}{4-x} \cdot \dfrac{8x-32}{x-2} = \dfrac{3(x^2-4)}{4-x} \cdot \dfrac{8(x-4)}{x-2}$

$= \dfrac{3(x+2)(x-2)}{4-x} \cdot \dfrac{-8(4-x)}{x-2} = \dfrac{3(x+2)\,\cancel{(x-2)}}{\cancel{4-x}} \cdot \dfrac{-8\,\cancel{(4-x)}}{\cancel{x-2}}$

$= -24(x+2)$

9.6 $\dfrac{x^2-2x-3}{x^2-x-6} \div \dfrac{x^2+3x+2}{x^2+4x+4} = \dfrac{x^2-2x-3}{x^2-x-6} \cdot \dfrac{x^2+4x+4}{x^2+3x+2}$

$= \dfrac{\cancel{(x-3)}\,\cancel{(x+1)}}{\cancel{(x-3)}\,\cancel{(x+2)}} \cdot \dfrac{\cancel{(x+2)}\,\cancel{(x+2)}}{\cancel{(x+2)}\,\cancel{(x+1)}} = 1$

9.7 Common denominator: x^2y;

$\dfrac{x+1}{x^2y} + \dfrac{2y}{x} \cdot \dfrac{xy}{xy} = \dfrac{x+1}{x^2y} + \dfrac{2xy^2}{x^2y} = \dfrac{2xy^2+x+1}{x^2y}$

9.8   Common denominator: $(x-3)(x+2)$

$$\frac{x+12}{(x-3)(x+2)} - \frac{x}{x+2} = \frac{x+12}{(x-3)(x+2)} - \frac{x}{x+2} \cdot \frac{x-3}{x-3}$$

$$= \frac{x+12}{(x-3)(x+2)} - \frac{x^2-3x}{(x-3)(x+2)} = \frac{x+12-(x^2-3x)}{(x-3)(x+2)} = \frac{-x^2+4x+12}{(x-3)(x+2)}$$

$$= \frac{-(x^2-4x-12)}{(x-3)(x+2)} = \frac{-(x-6)\,\overset{1}{\cancel{(x+2)}}}{(x-3)\,\cancel{(x+2)}} = \frac{-(x-6)}{x-3} \text{ or } \frac{6-x}{x-3}$$

9.9 $$\frac{\dfrac{x}{y}}{\dfrac{3}{y-1}} = \frac{x}{y} \div \frac{3}{y-1} = \frac{x}{y} \cdot \frac{y-1}{3} = \frac{xy-x}{3y}$$

9.10   Common denominator: x^2

$$\frac{x+\dfrac{1}{x}}{2-\dfrac{1}{x^2}} \cdot \frac{x^2}{x^2} = \frac{x^2\left(x+\dfrac{1}{x}\right)}{x^2\left(2-\dfrac{1}{x^2}\right)} = \frac{x^2(x)+x^2\left(\dfrac{1}{x}\right)}{x^2(2)-x^2\left(\dfrac{1}{x^2}\right)} = \frac{x^3+x}{2x^2-1}$$

9.11   Restricted value: 0; Common denominator: $6x$

$$6x\left(\frac{x+5}{6} - 4\right) = \left(\frac{-3}{x}\right)6x; \ \overset{1}{\cancel{6}}x\left(\frac{x+5}{\cancel{6}_{1}}\right) + 6x(-4) = \left(\frac{-3}{\cancel{x}}\right)6\overset{1}{\cancel{x}};$$

$$x^2+5x-24x = -18; \ x^2-19x+18 = 0; \ (x-18)(x-1) = 0;$$

$$x = 18 \text{ or } 1; \text{ Check } x = 18: \frac{18+5}{6} - 4 = \frac{-3}{18}; \ \frac{23}{6} - \frac{24}{6} = \frac{-1}{6} \checkmark;$$

$$x = 1: \frac{1+5}{6} - 4 = \frac{-3}{1}; \ 1 - 4 = -3 \checkmark$$

9.12 $x + 4 = 0$ when $x = -4$; $x - 2 = 0$ when $x = 2$; $x - 1 = 0$ when $x = 1$; Restricted values: $-4, 2, 1$
Common denominator: $(x + 4)(x - 2)(x - 1)$

$$(x+4)(x-2)(x-1)\left(\frac{1}{x+4} - \frac{2}{x-2}\right) = \left(\frac{-3}{x-1}\right)(x+4)(x-2)(x-1);$$

$$(x-2)(x-1) - 2(x+4)(x-1) = -3(x+4)(x-2);$$

$$x^2 - 3x + 2 - 2x^2 - 6x + 8 = -3x^2 - 6x + 24;$$

$$-x^2 - 9x + 10 = -3x^2 - 6x + 24; \quad 2x^2 - 3x - 14 = 0;$$

$$(2x - 7)(x + 2) = 0; \quad x = \frac{7}{2} \text{ or } -2; \text{ Check } x = \frac{7}{2}:$$

$$\frac{1}{\frac{7}{2}+4} - \frac{2}{\frac{7}{2}-2} = \frac{-3}{\frac{7}{2}-1}; \quad \frac{1}{2}\left(\frac{1}{\frac{7}{2}+4} - \frac{2}{\frac{7}{2}-2}\right) = \left(\frac{-3}{\frac{7}{2}-1}\right)\frac{1}{2};$$

$$\frac{1}{2\left(\frac{7}{2}+4\right)} - \frac{2}{2\left(\frac{7}{2}-2\right)} = \frac{-3}{\left(\frac{7}{2}-1\right)2}; \quad \frac{1}{15} - \frac{2}{3} = \frac{-3}{5}; \quad \frac{1}{15} - \frac{10}{15} = \frac{-9}{15} \checkmark$$

$$x = -2: \quad \frac{1}{-2+4} - \frac{2}{-2-2} = \frac{-3}{-2-1}; \quad \frac{1}{2} + \frac{1}{2} = 1 \checkmark$$

9.13   If $x = -3$, then $x + 3 = 0$, so $x = -3$ is a restricted value. The domain of $f(x)$ is all real numbers except -3.

$$f(x) = \frac{x^2 - x - 12}{x + 3} = \frac{(x-4)\overset{1}{\cancel{(x+3)}}}{\underset{1}{\cancel{x+3}}} = x - 4$$

$f(x) = x - 4$ is a linear function. The graph is a line with a slope of 1 and a y-intercept of -4. Since the domain does not include -3, x cannot equal -3 and the point $(-3, -7)$ is excluded from the graph.

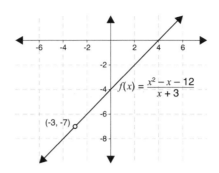

9.14 $x = 0$ is a restricted value. The domain of $f(x)$ is all real numbers except 0.

$$f(x) = \frac{x^2 - x}{x} = \frac{\overset{1}{\cancel{x}}(x-1)}{\underset{1}{\cancel{x}}} = x - 1; f(x) = x - 1 \text{ is a linear function.}$$

The graph is a line with a slope of 1 and a y-intercept of -1. Since the domain does not include 0, x cannot equal 0 and the point $(0, -1)$ is excluded from the graph.

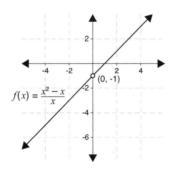

9.15   If $x = 3$, then $x - 3 = 0$, so $x = 3$ is a restricted value. The domain of $f(x)$ is all real numbers except 3. The function cannot be simplified. If $x - 3 = 0$, then $x = 3$, so $x = 3$ is a vertical asymptote. $p(x) = 6x + 2$, so the degree of $p(x) = 1$. $q(x) = x - 3$, so the degree of $q(x) = 1$. The degree of $p(x) = $ the degree of $q(x)$, so

$y = \dfrac{6}{1} = 6$ is a horizontal asymptote. Choose values in each

region to graph: $f(x) = \dfrac{6x + 2}{x - 3}$

x	−12	−2	0	1	2
f(x)	4.67	2	−0.67	−4	−14
x	4	6	8	12	20
f(x)	26	12.67	10	8.22	7.18

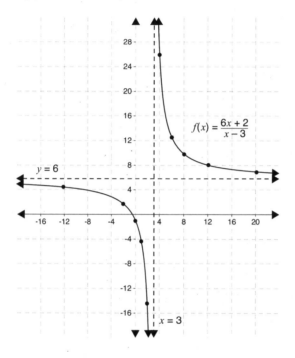

$$f(x) = \frac{6x + 2}{x - 3}$$

9.16   If $x = 5$, then $x - 5 = 0$, so $x = 5$ is a restricted value. The domain of $f(x)$ is all real numbers except 5. The function cannot be simplified. If $x - 5 = 0$, then $x = 5$, so $x = 5$ is a vertical asymptote. $p(x) = -2$, so the degree of $p(x) = 0$. $q(x) = x - 5$, so the degree of $q(x) = 1$. The degree of $p(x) <$ the degree of $q(x)$, so $y = 0$ is a horizontal asymptote. Choose values in each region to graph:

$$f(x) = \frac{-2}{x - 5}$$

x	−5	0	3	4.5	4.75
f(x)	0.2	0.4	1	4	8
x	5.25	5.5	6	7	10
f(x)	−8	−4	−2	−1	−0.4

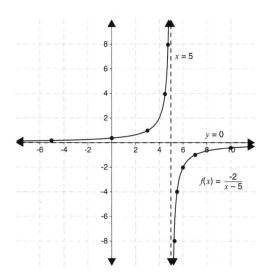

9.17   If $x = 3$, then $x - 3 = 0$, so $x = 3$ is a restricted value. The domain of $f(x)$ is all real numbers except 3. The function cannot be simplified. If $x - 3 = 0$, then $x = 3$, so $x = 3$ is a vertical asymptote. $p(x) = 2x^2$, so the degree of $p(x) = 2$. $q(x) = x - 3$ so the degree of $q(x) = 1$. The degree of $p(x)$ is one more than the degree of $q(x)$, so the function has an oblique asymptote.

$$x - 3 \overline{)\,2x^2}^{\displaystyle 2x + 6}$$

$y = 2x + 6$ is the oblique asymptote.

$$\underline{-\left(2x^2 - 6x\right)}$$
$$6x$$
$$\underline{-\left(6x - 18\right)}$$
$$18$$

Choose values to graph: $f(x) = \dfrac{2x^2}{x - 3}$

x	−7	−5	0	2	2.25
f(x)	−9.8	−6.25	0	−8	−13.5
x	3.75	4	6	9	12
f(x)	37.5	32	24	27	32

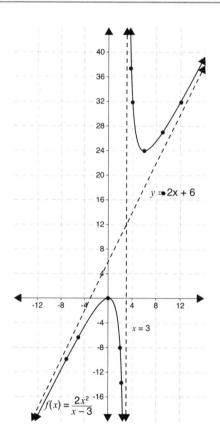

$y = 2x + 6$

$x = 3$

$f(x) = \dfrac{2x^2}{x-3}$

9.18 $f(x) = \dfrac{x^3 + 3x^2}{x^2 + 5x + 6} = \dfrac{x^2(x+3)}{(x+3)(x+2)}$

If $x = -3$, then $(x + 3)(x + 2) = 0$, so $x = -3$ is a restricted value.
If $x = -2$, then $(x + 3)(x + 2) = 0$, so $x = -2$ is a restricted value.
The domain of $f(x)$ is all real numbers except -3 and -2.

$$f(x) = \dfrac{x^3 + 3x^2}{x^2 + 5x + 6} = \dfrac{x^2 \, \overset{1}{\cancel{(x+3)}}}{\underset{1}{\cancel{(x+3)}}\,(x+2)} = \dfrac{x^2}{x+2} \; ;$$

If $x + 2 = 0$, then $x = -2$, so $x = -2$ is a vertical asymptote.
$p(x) = x^2$, so the degree of $p(x) = 2$. $q(x) = x + 2$ so the degree of
$q(x) = 1$. The degree of $p(x)$ is one more than the degree of $q(x)$,
so the function has an oblique asymptote.

$$\begin{array}{r} x-2 \\ x+2\overline{\smash{\big)}\,x^2 } \\ -\left(x^2+2x\right) \\ \hline -2x \\ -\left(-2x+4\right) \\ \hline -4 \end{array}$$

$y = x - 2$ is the oblique asymptote.

Choose values to graph: $f(x) = \dfrac{x^2}{x+2}$

x	-10	-6	-4	-3	-2.5
$f(x)$	-12.5	-9	-8	-9	-12.5
x	-1.5	-1	0	3	6
$f(x)$	4.5	1	0	1.8	4.5

Recall that -3 is restricted from the domain, so there is a point discontinuity at $(-3, 9)$.

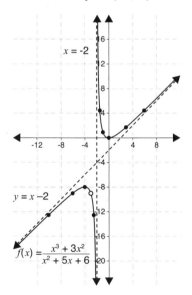

9.19 $y = kx;\ 7 = k(14);\ \dfrac{1}{2} = k;\ y = \dfrac{1}{2}x$

9.20 $\dfrac{y_1}{x_1} = \dfrac{y_2}{x_2};\ \dfrac{8}{4} = \dfrac{9}{x_2};\ 8x_2 = 36;\ x_2 = \dfrac{36}{8} = \dfrac{9}{2}$

9.21 $y = \dfrac{k}{x}$; $8 = \dfrac{k}{6}$; $48 = k$; $y = \dfrac{48}{x}$

9.22 $x_1 y_1 = x_2 y_2$; $2(3) = x_2 \cdot 5$; $6 = 5x_2$; $\dfrac{6}{5} = x_2$

9.23 $y = kxz$; $72 = k(9)(4)$; $2 = k$; $y = 2xz$

9.24 $\dfrac{y_1}{x_1 z_1} = \dfrac{y_2}{x_2 z_2}$; $\dfrac{15}{3(4)} = \dfrac{8}{x_2(2)}$; $\dfrac{5}{4} = \dfrac{4}{x_2}$; $5x_2 = 16$; $x_2 = \dfrac{16}{5}$

9.25   Direct variation: $y = km$; Inverse variation: $y = \dfrac{k}{n^2}$

   Combined variation: $y = \dfrac{km}{n^2}$; $6 = \dfrac{k \cdot 4}{2^2}$; $6 = k$; $y = \dfrac{6m}{n^2}$

9.26 $y = \dfrac{6m}{n^2}$; $y = \dfrac{6 \cdot 7}{3^2}$; $y = \dfrac{14}{3}$

Chapter 10

10.1 $6 - 3 = 3$; $9 - 6 = 3$; $12 - 9 = 3$; $16 - 12 = 4$. There is not a common difference. The sequence is not arithmetic.

10.2 $-5 - (-3) = -2$; $-7 - (-5) = -2$; $-9 - (-7) = -2$; $-11 - (-9) = -2$ The common difference is -2. The sequence is an arithmetic sequence. $-11 + (-2) = -13$; $-13 + (-2) = -15$; $-15 + (-2) = -17$; $-17 + (-2) = -19$. The next four terms are $-13, -15, -17$, and -19.

10.3 $\dfrac{-6}{2} = -3$; $\dfrac{18}{-6} = -3$; $\dfrac{-54}{18} = -3$; $\dfrac{162}{-54} = -3$. The common ratio is -3.

   The sequence is a geometric sequence. $162 \cdot -3 = -486$;

   $-486 \cdot -3 = 1{,}458$; $1{,}458 \cdot -3 = -4{,}374$; $-4{,}374 \cdot -3 = 13{,}122$ The next four terms are $-486, 1{,}458, -4{,}374$, and $13{,}122$.

10.4 $\dfrac{6}{3} = 2$; $\dfrac{9}{6} = \dfrac{3}{2}$. There is no common ratio. The sequence is not geometric.

10.5 $8 - 6 = 2$; $11 - 8 = 3$. There is not a common difference.

The sequence is not arithmetic. $\frac{8}{6} = \frac{4}{3}$; $\frac{11}{8} = \frac{11}{8}$. There is not a common ratio. The sequence is not geometric. The sequence is neither arithmetic nor geometric.

10.6 $1 - 5 = -4$, $\frac{1}{5} - 1 = -\frac{4}{5}$. There is not a common difference.

The sequence is not arithmetic. $\frac{1}{5} = \frac{1}{5}$; $\frac{\frac{1}{5}}{1} = \frac{1}{5}$; $\frac{\frac{1}{25}}{\frac{1}{5}} = \frac{1}{25} \cdot \frac{5}{1} = \frac{1}{5}$;

$\frac{\frac{1}{125}}{\frac{1}{25}} = \frac{1}{125} \cdot \frac{25}{1} = \frac{1}{5}$; There is a common ratio.

The sequence is geometric.

10.7 $-5 - (-9) = 4$, $-1 - (-5) = 4$; $3 - (-1) = 4$; $7 - 3 = 4$. There is a common difference. The sequence is arithmetic.

10.8 Common difference $= 2 - 8 = -6$; $d = -6$, $n = 100$, $a_1 = 8$;

$a_n = d(n-1) + a_1$; $a_{100} = -6(100-1) + 8$; $a_{100} = -6(99) + 8$;

$a_{100} = -586$. The 100th term of the sequence is -586.

10.9 $a_{10} = 59$, $n = 10$, $d = 6$; $a_n = d(n-1) + a_1$; $59 = 6(10-1) + a_1$;

$59 = 54 + a_1$; $5 = a_1$; $a_1 = 5$, $d = 6$, $n = 50$; $a_n = d(n-1) + a_1$;

$a_{50} = 6(50-1) + 5$; $a_{50} = 299$. The 50th term is 299.

10.10 $a_1 = 4$, $a_2, a_3, a_4, a_5 = -8$; $a_n = d(n-1) + a_1$; $-8 = d(5-1) + 4$

$-8 = 4d + 4$; $-12 = 4d$–$3 = d$; $4 + (-3) = 1$; $1 + (-3) = -2$,

$-2 + (-3) = -5$; The arithmetic means are 1, -2 and -5.

10.11   Common difference $= -6 - (-10) = 4$; $a_1 = -10$, $d = 4$, $n = 10$

$$S_n = \frac{n}{2}\left[2a_1 + (n-1)d\right]; \ S_{10} = \frac{10}{2}\left[2(-10) + (10-1)4\right];$$

$$S_{10} = 5\left[-20 + (9)4\right]; \ S_{10} = 80$$

10.12 $a_4 = 6$, $a_9 = 16$; $a_n = d(n-1) + a_1$; $6 = d(4-1) + a_1$; $6 = 3d + a_1$;

$6 - 3d = a_1$; $a_n = d(n-1) + a_1$; $16 = d(9-1) + a_1$; $16 = 8d + a_1$;

$16 = 8d + (6 - 3d)$; $16 = 5d + 6$; $10 = 5d$; $2 = d$; $6 - 3d = a_1$;

$6 - 3(2) = a_1$; $0 = a_1$; $a_1 = 0$, $d = 2$, $n = 50$; $S_n = \frac{n}{2}\left[2a_1 + (n-1)d\right]$;

$$S_{50} = \frac{50}{2}\left[2(0) + (50-1)2\right]; \ S_{50} = 25\left[0 + 98\right]; \ S_{50} = 2{,}450$$

10.13   Common ratio $= \dfrac{1{,}024}{2{,}048} = \dfrac{1}{2}$; $r = \dfrac{1}{2}$, $a_1 = 2{,}048$, $n = 12$

$$a_n = a_1 r^{n-1}; \ a_{12} = 2{,}048\left(\frac{1}{2}\right)^{12-1}; \ a_{12} = 2{,}048\left(\frac{1}{2}\right)^{11}; \ a_{12} = 2{,}048 \cdot \frac{1}{2^{11}};$$

$a_{12} = 2{,}048 \cdot \dfrac{1}{2{,}048} = 1$. The 12th term of the sequence is 1.

10.14 $a_6 = 256$, $r = -2$; $a_n = a_1 r^{n-1}$; $256 = a_1(-2)^{6-1}$; $256 = a_1(-2)^5$;

$256 = a_1(-32)$; $-8 = a_1$; $a_1 = -8$, $r = -2$, $n = 20$; $a_n = a_1 r^{n-1}$;

$a_{20} = -8(-2)^{20-1}$; $a_{20} = -8(-2)^{19}$; $a_{20} = 4{,}194{,}304$

The 20th term is 4,194,304.

10.15 $a_1 = 5$, a_2, a_3, a_4, a_5, $a_6 = 1{,}215$, $a_1 = 5$, $a_6 = 1{,}215$, and $n = 6$;

$a_n = a_1 r^{n-1}$; $1{,}215 = 5r^{6-1}$; $1{,}215 = 5r^5$; $243 = r^5$; $(243)^{\frac{1}{5}} = \left(r^5\right)^{\frac{1}{5}}$;

$3 = r$; $5(3) = 15$; $15(3) = 45$; $45(3) = 135$, $135(3) = 405$

The geometric means are 15, 45, 135, and 405.

10.16 $a_1 = -2$, $r = \dfrac{1}{3}$; $S_n = \dfrac{a_1 - a_1 r^n}{1-r}$; $S_8 = \dfrac{-2 - (-2)\left(\dfrac{1}{3}\right)^8}{1 - \left(\dfrac{1}{3}\right)}$;

$S_8 = \dfrac{-2 + 2\left(\dfrac{1}{3}\right)^8}{\dfrac{2}{3}}$; $S_8 = \left(-2 + 2\left(\dfrac{1}{3}\right)^8\right) \cdot \dfrac{3}{2}$; $S_8 = \left(-1 + 1\left(\dfrac{1}{3}\right)^8\right) \cdot 3$;

$S_8 = \left(-1 + \dfrac{1}{6{,}561}\right) \cdot 3$; $S_8 \approx -2.99954$

10.17 $a_1 = 96$ and $a_6 = 3{,}072$; $a_n = a_1 r^{n-1}$; $3{,}072 = 96 r^{6-1}$; $3{,}072 = 96 r^5$;

$\dfrac{3{,}072}{96} = r^5$; $32 = r^5$; $\left(2\right)^5 = r^5$; $2 = r$; $a_7 = a_6 \cdot 2 = 3{,}072 \cdot 2 = 6{,}144$;

$a_1 = 96$, $a_7 = 6{,}144$, $n = 7$, $r = 2$; $S_n = \dfrac{a_1 - a_n r}{1-r}$;

$S_7 = \dfrac{96 - 6{,}144(2)}{1 - 2}$; $S_7 = \dfrac{96 - 12{,}288}{-1}$; $S_7 = 12{,}192$

10.18 $\displaystyle\sum_{m=0}^{5} -6m = -6(0) + (-6)(1) + (-6)(2) + (-6)(3) + (-6)(4) + (-6)(5)$

$= 0 + (-6) + (-12) + (-18) + (-24) + (-30) = -90$

10.19 When $k = 1$, $2\left(\dfrac{1}{2}\right)^{k-1} = 2\left(\dfrac{1}{2}\right)^{1-1} = 2\left(\dfrac{1}{2}\right)^0 = 2(1) = 2 = a_1$

When $k = 6$, $2\left(\dfrac{1}{2}\right)^{k-1} = 2\left(\dfrac{1}{2}\right)^{6-1} = 2\left(\dfrac{1}{2}\right)^5 = 2\left(\dfrac{1}{32}\right) = \dfrac{1}{16} = a_6$

$a_1 = 2$, $a_6 = \dfrac{1}{16}$, $r = \dfrac{1}{2}$, $n = 6$; $S_n = \dfrac{a_1 - a_n r}{1-r}$; $S_6 = \dfrac{2 - \dfrac{1}{16}\left(\dfrac{1}{2}\right)}{1 - \dfrac{1}{2}}$;

$$S_6 = \frac{2 - \frac{1}{32}}{\frac{1}{2}}; \; S_6 = \left(2 - \frac{1}{32}\right) \cdot 2 = 4 - \frac{1}{16} = \frac{63}{16}$$

10.20 $12 - 7 + 1 = 6$ terms. When $t = 7$, $-2t + 6 = -2(7) + 6 = -8 = a_1$.

When $t = 12$, $-2t + 6 = -2(12) + 6 = -18 = a_6$. $S_n = \frac{n}{2}(a_1 + a_n)$;

$S_6 = \frac{6}{2}(-8 + (-18))$; $S_6 = 3(-26)$; $S_6 = -78$

10.21   There are five terms, so $n = 5$; $8 - 5 = 3$; $11 - 8 = 3$; $14 - 11 = 3$;

$17 - 14 = 3$. The series is an arithmetic series with a common

difference of 3. $a_1 = 5$, $d = 3$; $a_n = d(n-1) + a_1$; $a_n = 3(n-1) + 5$;

$a_n = 3n - 3 + 5$; $a_n = 3n + 2$; $\displaystyle\sum_{n=1}^{5} 3n + 2$

10.22   a. $r = \frac{5}{1} = 5$; $|r| = 5 \geq 1$. The series diverges.

b. $r = \frac{54}{81} = \frac{1}{3}$; $|r| = \frac{1}{3} < 1$. The series converges.

10.23   a. $r = \frac{12}{6} = 2$; $|r| = 2 \geq 1$. The series diverges and has no sum.

b. $r = \frac{-9}{-18} = \frac{1}{2}$; $|r| = \frac{1}{2} < 1$. The series converges.

$$S = \frac{a_1}{1 - r}; \; a_1 = -18, \; r = \frac{1}{2}; \; S = \frac{-18}{1 - \frac{1}{2}} = \frac{-18}{\frac{1}{2}} = -18 \cdot \frac{2}{1} = -36$$

10.24 $|r| = \frac{2}{3} < 1$, so the series converges.

$$a_1 = 6\left(\frac{2}{3}\right)^3 = 6\left(\frac{8}{27}\right) = \frac{16}{9}; \; S = \frac{a_1}{1 - r}; \; S = \frac{\frac{16}{9}}{1 - \frac{2}{3}} = \frac{\frac{16}{9}}{\frac{1}{3}} = \frac{16}{9} \cdot \frac{3}{1} = \frac{16}{3}$$

The sum of the series is $\dfrac{16}{3}$.

10.25 The solution to the first example of the lesson shows the 7th row of Pascal's triangle.

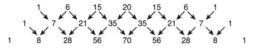

10.26 $7! = 7 \cdot 6 \cdot 5 \cdot 4 \cdot 3 \cdot 2 \cdot 1 = 5{,}040$

10.27 In the first term, the exponent of a is 12. In the second term, the exponent of a is 11. In the third term, the exponent of a is 10. In the fourth term, the exponent of a is 9. In the fifth term, the exponent of a is 8. In the sixth term, the exponent of a is 7. In the seventh term, the exponent of a is 6 and the exponent of b is $12 - 6 = 6$.

$$\frac{n!}{(\text{exponent of }a)!(\text{exponent of }b)!} = \frac{12!}{6!6!}$$

$$= \frac{\overbrace{12 \cdot 11 \cdot 10 \cdot 9 \cdot 8 \cdot 7 \cdot 6 \cdot 5 \cdot 4 \cdot 3 \cdot 2 \cdot 1}^{12!}}{\underbrace{(6 \cdot 5 \cdot 4 \cdot 3 \cdot 2 \cdot 1)}_{6!}\underbrace{(6 \cdot 5 \cdot 4 \cdot 3 \cdot 2 \cdot 1)}_{6!}}$$

$$= \frac{\cancel{12}^{\,2} \cdot 11 \cdot \cancel{10} \cdot \cancel{9}^{\,3} \cdot \cancel{8}^{\,4} \cdot 7 \cdot \cancel{6} \cdot \cancel{5} \cdot \cancel{4} \cdot \cancel{3} \cdot \cancel{2} \cdot 1}{(\cancel{6} \cdot \cancel{5} \cdot \cancel{4} \cdot \cancel{3} \cdot \cancel{2} \cdot 1)(\cancel{6} \cdot \cancel{5} \cdot \cancel{4} \cdot \cancel{3} \cdot \cancel{2}_{2} \cdot 1)} = 11 \cdot 3 \cdot 4 \cdot 7 = 924;$$

The seventh term is $924a^6 b^6$.

10.28 $(a+b)^n = a^n + \dfrac{n}{1}a^{n-1}b + \dfrac{n(n-1)}{2 \cdot 1}a^{n-2}b^2 + \dfrac{n(n-1)(n-2)}{3 \cdot 2 \cdot 1}a^{n-3}b^3 + \ldots + b^n$

$(x+y)^6 = x^6 + \dfrac{6}{1}x^5 y + \dfrac{6(5)}{2 \cdot 1}x^4 y^2 + \dfrac{6(5)(4)}{3 \cdot 2 \cdot 1}x^3 y^3 + \dfrac{6(5)(4)(3)}{4 \cdot 3 \cdot 2 \cdot 1}x^2 y^4 +$

$\dfrac{6(5)(4)(3)(2)}{5 \cdot 4 \cdot 3 \cdot 2 \cdot 1}xy^5 + y^6$

$= x^6 + 6x^5 y + 15x^4 y^2 + 20x^3 y^3 + 15x^2 y^4 + 6xy^5 + y^6$

10.29 $(a+b)^n = a^n + \dfrac{n}{1}a^{n-1}b + \dfrac{n(n-1)}{2\cdot 1}a^{n-2}b^2 + \dfrac{n(n-1)(n-2)}{3\cdot 2\cdot 1}a^{n-3}b^3 + \ldots + b^n$

$(3x+1)^4 = (3x)^4 + \dfrac{4}{1}(3x)^3 1 + \dfrac{4(3)}{2\cdot 1}(3x)^2 1^2 + \dfrac{4(3)(2)}{3\cdot 2\cdot 1}(3x)1^3 + 1^4$

$= 81x^4 + 4(27)x^3 + 6(9)x^2 + 4(3)x + 1 = 81x^4 + 108x^3 + 54x^2 + 12x + 1$